Analog Circuits and Signal Processing

The *Analog Circuits and Signal Processing* book series, formerly known as the *Kluwer International Series in Engineering and Computer Science*, is a high level academic and professional series publishing research on the design and applications of analog integrated circuits and signal processing circuits and systems. Typically per year we publish between 5-15 research monographs, professional books, handbooks, edited volumes and textbooks with worldwide distribution to engineers, researchers, educators, and libraries.

The book series promotes and expedites the dissemination of new research results and tutorial views in the analog field. There is an exciting and large volume of research activity in the field worldwide. Researchers are striving to bridge the gap between classical analog work and recent advances in very large scale integration (VLSI) technologies with improved analog capabilities. Analog VLSI has been recognized as a major technology for future information processing. Analog work is showing signs of dramatic changes with emphasis on interdisciplinary research efforts combining device/circuit/technology issues. Consequently, new design concepts, strategies and design tools are being unveiled.

Topics of interest include:
Analog Interface Circuits and Systems;
Data converters; Active-RC, switched-capacitor and continuous-time integrated filters;
Mixed analog/digital VLSI;
Simulation and modeling, mixed-mode simulation;
Analog nonlinear and computational circuits and signal processing;
Analog Artificial Neural Networks/Artificial Intelligence;
Current-mode Signal Processing; Computer-Aided Design (CAD) tools;
Analog Design in emerging technologies (Scalable CMOS, BiCMOS, GaAs, heterojunction and floating gate technologies, etc.);
Analog Design for Test;
Integrated sensors and actuators; Analog Design Automation/Knowledge-based Systems; Analog VLSI cell libraries; Analog product development; RF Front ends, Wireless communications and Microwave Circuits;
Analog behavioral modeling, Analog HDL.

More information about this series at http://www.springer.com/series/7381

Mohammad Elbadry • Ramesh Harjani

Quadrature Frequency Generation for Wideband Wireless Applications

Mohammad Elbadry
Qualcomm Technologies Inc.
San Diego
California
USA

Ramesh Harjani
Department of ECE
University of Minnesota
Minneapolis
Minnesota
USA

ISSN 1872-082X ISSN 2197-1854 (electronic)
Analog Circuits and Signal Processing
ISBN 978-3-319-38516-7 ISBN 978-3-319-13788-9 (eBook)
DOI 10.1007/978-3-319-13788-9

Springer Cham Heidelberg New York Dordrecht London

Softcover reprint of the hardcover 1st edition 2015

Printed on acid-free paper

Springer is part of Springer Science+Business Media (www.springer.com)

To my dear parents. . .

Preface

This book provides an overview on quadrature generation techniques for wireless transceivers. The main focus is on LC quadrature VCO (QVCO) design. The book tries to provide a simple and intuitive understanding of the operation of quadrature LC VCOs, as well as a discussion of their drawbacks and limitations. The discussion is provided from a designer's perspective, allowing a simple understanding. Based on this understanding, a general classification of the quadrature LC VCO techniques in literature is provided, together with the advantages and limitations of each. Building on this understanding, two new quadrature coupling techniques are introduced. The first technique allows robust quadrature coupling with reduced phase-noise, by using a simple modification of the conventional QVCO. The coupling scheme doesn't require any frequency-sensitive coupling networks, or complex transformer-based designs. The second technique is targeted at millimeter-wave applications, allowing a wide-tuning range QVCO design at 60 GHz and beyond. This is all culminated by the introduction of a frequency-synthesis scheme for a 4 GHz instantaneous bandwidth channelized receiver. Injection locking techniques are used, together with QVCOs, to generate two simultaneous quadrature LOs at 20 GHz and 22 GHz using a single-ended, single-phase, 1.33 GHz reference clock.

Contents

List of Figures

List of Tables

Chapter 1
Wideband Transceivers: Challenges

With the proliferation of wireless technology in the last two decades, a plethora of new devices and applications have revolutionized our everyday life. Portable devices with wireless connectivity have become ubiquitous, while continuously getting thinner and lighter (Fig. 1.1). Moreover, wireless data rates have increased rapidly from the slow rates of the earliest cellular networks of the late nineties to the dazzling speeds of the next generation 4G-LTE (Fig. 1.2). In addition to the increased speeds, the wireless capabilities have also increased. Modern smartphones and tablets have GPS (for navigation), Bluetooth (for wireless headsets and file transfer), WiFi (for wireless internet), GSM and CDMA (for phone calls and texting), and FM radio. And some even have wireless NFC (Near Field Communication) capabilities allowing us to pay for drinks and food on the go with cell phones.

These advances, however, do not come for free. Higher data rates come with the cost of higher power consumption, adversely affecting battery life of portable devices. This is further aggravated by the small form factors which don't allow large batteries. Additionally, the multiple radios needed require the use of a large number of integrated circuit chips. This, in turn, makes it more difficult to reduce form factors and power consumption. With the widespread use of high-definition multi-media, data rates are projected to increase and higher speed wireless standards (such as WirelessHD [3]) are expected to emerge.

In high data rate wireless receivers, a major power hog is the analog-to-digital converter (ADC) (which is responsible for converting radio signals at the antenna into digital ones-and-zeros that can be shown on the screen or heard via the speaker). One way to reduce the power consumption in such receivers is to split the incoming signal into a number of parallel streams (channelization), each with a smaller data rate. The aggregate data rate remains the same, but the overall energy is reduced [4]. On the other hand, the number of chips can be reduced by incorporating programmability into the radio design, allowing a single radio to be used with multiple wireless standards, i.e., by designing software defined radios (SDR). An important component of an SDR is wide tuning range quadrature oscillator, that enables quadrature LO generation for direct downconversion in different bands.

M. Elbadry, R. Harjani, *Quadrature Frequency Generation for Wideband Wireless Applications,* Analog Circuits and Signal Processing, DOI 10.1007/978-3-319-13788-9_1

Fig. 1.1 Evolution of the cell-phone [1]

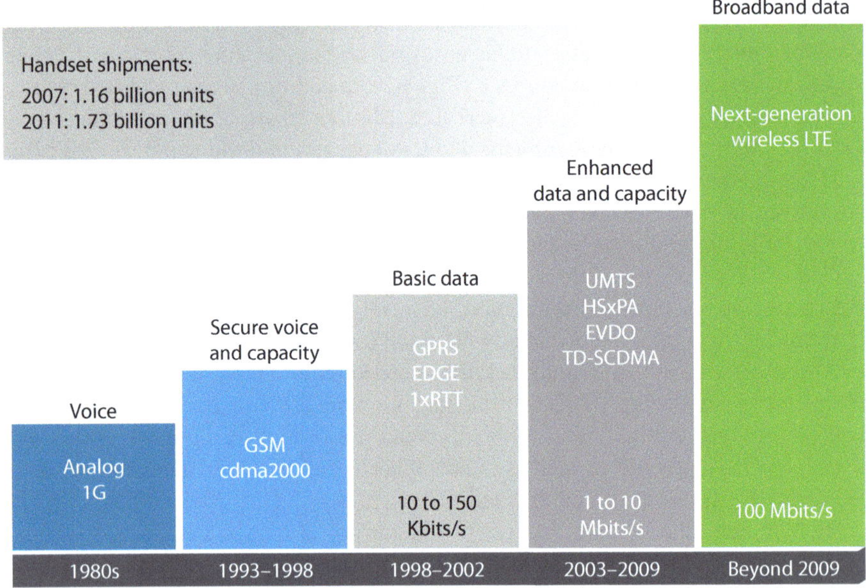

Fig. 1.2 Evolution of the cellular data rates [2]

1.1 Software Defined Radio

Software-defined radio receivers allow a single front-end to be used for multiple standards, through software programmability and reconfigurability. The original "Mitola" [5] SDR has an ADC directly following the antenna, as shown in Fig. 1.3, allowing all downconversion and post-processing to be performed in digital domain.

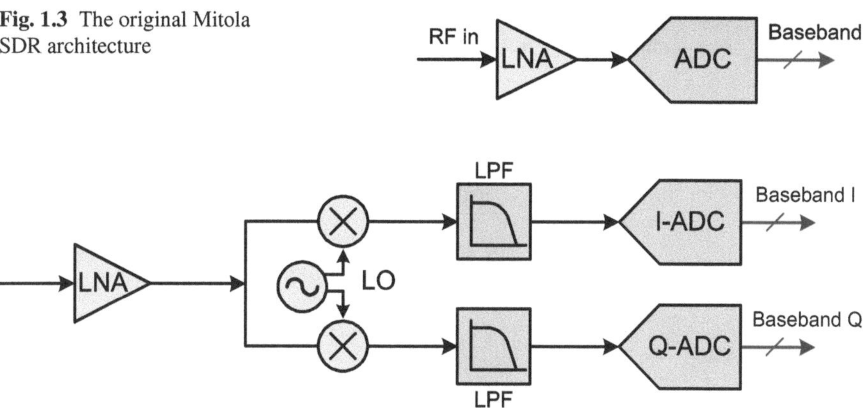

Fig. 1.3 The original Mitola SDR architecture

Fig. 1.4 A typical direct-conversion receiver

This approach is not practical with current technology due to the high power consumption needed for the front-end LNA as well as the the very high clocking speed and very high linearity requirements on the ADC.

A more practical approach is to use a generic direct-conversion receiver, similar to that shown in Fig. 1.4. To provide SDR functionality, all the blocks have to be programmable to tune the receiver to different bands. For the LO path, LC VCOs are usually used due to the strict phase-noise requirements of wireless communications. Achieving a very wideband (close to a decade) programmabiity is difficult for LC VCOs, thus requiring the use of banks of VCOs [6]. To generate quadrature phases, the most popular way is to operate the VCO at a higher frequency than the desired LO, and then use dividers to generate quadrature phases at the LO frequency. This incurs a power hit, due to the high power required for VCO buffering at the high frequency (without significant phase-noise degradation).

Quadrature VCOs provide an alternative method for the direct generation of quadrature LOs without requiring the VCO to operate at twice the desired operating frequency. For SDR applications, a wide tuning range in the QVCO is highly desirable as it allows less number of QVCOs in the bank. This, in turn, reduces the area requirements, complexity, and design time.

1.2 Frequency Channelization

Figure 1.4 shows a typical direct-conversion receiver. A low-noise amplifier amplifies the input signal, which is passed on to quadrature mixers for downconversion. Quadrature mixing is typically done by using quadrature LO, although quadrature signal generation can be done as well. A low-pass filter follows the mixer, to filter out the higher frequency component. Finally, an ADC converts the downconverted and filtered data into digital for demodulation and post-processing. A major burden of the receiver linearity lies on the ADC, since it is the last block in the receiver and

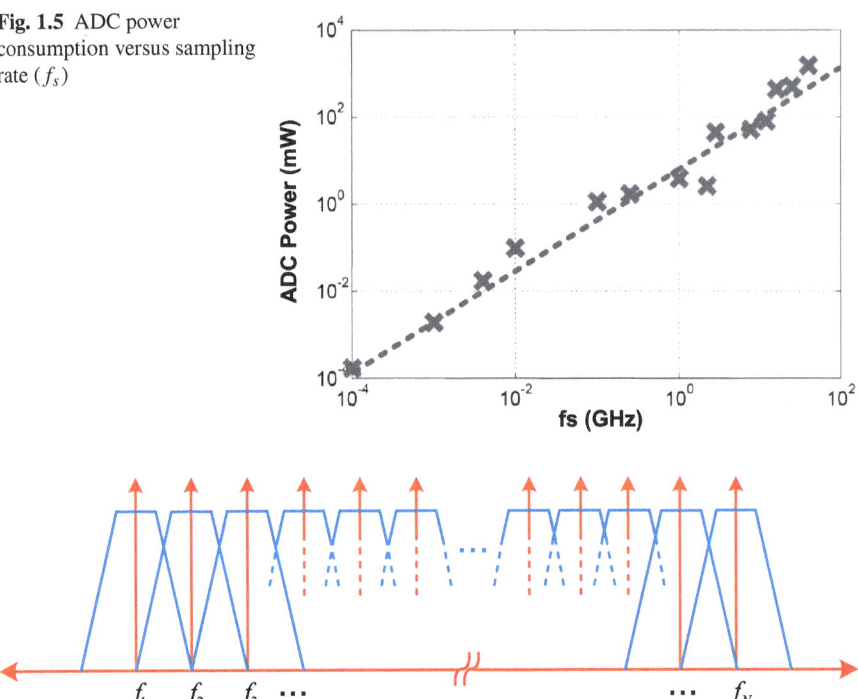

Fig. 1.5 ADC power consumption versus sampling rate (f_s)

Fig. 1.6 Frequency channelization

subject to large signal amplitudes specially in the presence of in-band blocker signals (the filter attenuates out-of-band blockers) [7]. Due to this linearity constraint, the ADC block usually has significant power consumption.

Unlike the RF front-end, the ADC power consumption rises exponentially with increased bandwidth. To illustrate this point, consider [8] which presents a GSM receiver with bandwidth of 200 kHz, and [9] which presents a 60 GHz WirelessHD receiver with 1 GHz of bandwidth. Although the bandwidth of the WirelessHD receiver in [9] is 50-times higher than the bandwidth of the GSM receiver in [8], the power consumption of the RF front-end in [9] is only 1.6 times higher than that in [8] (454 mW for WirelessHD versus 235 mW for GSM).

Figure 1.5, on the other hand, shows the power consumption of ADCs versus their sampling rate based on the data in [10]. At each sampling rate, the ADC with the lowest power consumption at that sampling rate is chosen for the plot. Power versus sampling frequency (f_s) is, then, plotted on a Log-Log scale. As evident from Fig. 1.5, the power consumption increases exponentially with sampling rate (f_s), i.e. $Power \propto (f_s)^n$. Fitting shows that $n \approx 2$; a large power saving can be achieved by reducing the ADC sampling rate.

To maintain the overall bandwidth, while reducing the ADC sampling rate, frequency channelization can be employed as shown in Fig. 1.6. The signal bandwidth

(BW) is divided into N different chunks, each of bandwidth $\frac{BW}{N}$. Hence, N different streams need to be digitized, requiring N ADCs each with a sampling rate $\frac{f_s}{N}$ (f_s being the sampling frequency of the unchannelized signal). Hence, the ratio of the power consumption of the channelized ADCs P_c to the power consumption of a single ADC P_s can be given by:

$$\frac{P_c}{P_s} \approx \frac{N \times \left(\frac{f_s}{N}\right)^2}{f_s^2} = \frac{1}{N} \tag{1.1}$$

This means that the ADC power consumption can be reduced roughly N times by using channelization. Actual power savings will depend on the extra power consumed for performing channelization, as well as additional amplification that might be needed post-channelization. Nevertheless, significant power savings can be achieved through channelization. Moreover, channelization improves the RF receiver performance by making the system more interference tolerant [11]. For instance, if an interferer falls onto one of the channels, that channel and the associated ADC can be shut-off, reducing the data rate but without compromising the overall performance. If a single ADC is used, on the other hand, a single large interferer can overwhelm leading to a total signal blockage. *A major challenge for channelization is the generation of multiple, uniformly spaced, quadrature LOs for downconversion of the different channels.*

1.3 Organization

This book is focused on quadrature LO generation techniques for wideband applications. "Wideband" includes instantaneous wide bandwidth system, as well as systems with smaller instantaneous bandwidth but have a wide range of center (or carrier frequencies).

With the QVCO being a major building block, Chap. 2 provides an overview of the current QVCO techniques used in literature and attempts to classify them. Based on this understanding, an improved QVCO scheme—which is also robust—is presented and discussed in Chap. 3. Also, a QVCO scheme suitable for mm-Wave applications (which allow wideband operation) is presented in Chap. 4.

Finally, QVCOs together wit injection locking are employed to implement an LO generation scheme for a channelized receiver; the details of which are presented in Chap. 5. This is followed by conclusions in Chap. 6.

Chapter 2
QVCO: Concepts and Classification

2.1 Quadrature VCOs: an Overview

With the advent of CMOS technology, direct downconversion receivers are gaining popularity due to low cost and simplicity. Quadrature LO generation is critical to the operation of direct-downconversion receivers [7]. Two common techniques for quadrature generation are divide-by-two frequency dividers, and polyphase filters [12]. Divide-by-two dividers require the system's oscillator to work at double the desired frequency, resulting in an overall increase in power consumption. Power consumption penalty is primarily due to the increase in power consumption of the VCO buffers. Polyphase filters allow quadrature generation without the need for doubling the frequency. The lossy nature of polyphase filters, however, results in increased power consumption (for buffering and amplifying signals). Moreover, polyphase quadrature accuracy is sensitive to absolute component values. Hence, multistage polyphase filters are often employed to combat process variations, resulting in increased losses [13]. Quadrature accuracy is also reduced if the polyphase filter input is not a pure sinusoid. Since inverter-based buffers are popular in today's submicron technologies, generating a pure sinusoid can be elusive. This introduces another source of quadrature error.

LC-based Quadrature Voltage Controlled Oscillators (QVCOs) allow the generation of quadrature LO signals without the need for doubling the frequency, and without the need for polyphase filtering. The basic structure of the LC-based QVCO is shown in Fig. 2.1 [14]. Two LC VCOs are coupled through both direct coupling (blue wires in Fig. 2.1), and cross coupling (red wires in Fig. 2.1). If both LC VCOs are matched, then owing to symmetry their differential outputs have to be in quadrature [15]. The QVCO can also be regarded as two inter-injection-locked VCOs [16].

The LC QVCO of Fig. 2.1 is very simple and robust. However, it has an inherent trade-off between phase-noise and quadrature accuracy. If we denote the coupling strength as α, then it can be defined as:

$$\alpha = \frac{I_{cp}}{I_{osc}} \tag{2.1}$$

M. Elbadry, R. Harjani, *Quadrature Frequency Generation for Wideband Wireless Applications*, Analog Circuits and Signal Processing, DOI 10.1007/978-3-319-13788-9_2

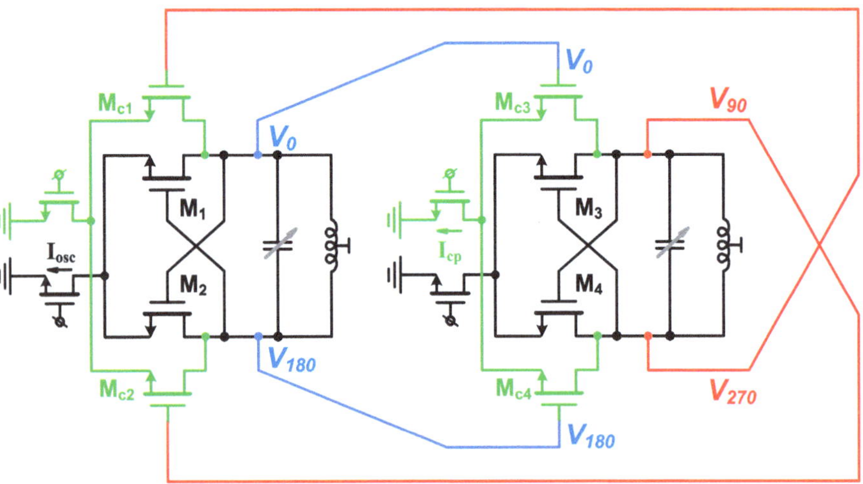

Fig. 2.1 The basic structure of an LC QVCO

where I_{cp} is the tail current of the coupling transistor pairs ($M_{c1}-M_{c2}$ and $M_{c3}-M_{c4}$), and I_{osc} is the tail current of the oscillator core transistor pairs (M_1-M_2 and M_3-M_4). Increasing the coupling strength (α), improves the quadrature accuracy, but leads to a degradation of phase noise, with increased phase noise contribution from the coupling transistors [12]. Hence, for reasonable quadrature accuracy, the LC QVCO of Fig. 2.1 has relatively poor phase-noise performance. Typically, for a given power dissipation the phase noise of the basic LC QVCO is 3–5 dB worse than a stand-alone oscillator [17].

Figure 2.2 shows one LC VCO from within the basic LC QVCO. The drain-source voltage (V_{ds}), gate-source voltage (V_{gs}) ,and drain current (I_{ds}) of the injection transistors are highlighted. The relatively poor phase-noise performance of the basic LC QVCO can be explained by referring to Fig. 2.3 which depicts the time-domain waveforms of V_{ds}, V_{gs} and I_{ds}.

Both V_{gs} and V_{ds} have a DC value of V_{dc}, which is lower than the supply voltage but considerably larger than the threshold voltage V_{th}. Both voltages swing with equal amplitudes around their DC value with a 90° phase shift. The peak value of the I_{ds} current, thus, occurs when V_{gs} is at its peak. Note that at this instant, V_{ds} is at its DC value (V_{dc}) which is relatively large. Hence, *the injection current (I_{ds}) has its peak value when the oscillator's output voltage ($V_{90}-V_{270}$) is at its zero-crossing*. Oscillators are most vulnerable to phase-noise when their output is zero crossing [18], hence this leads to a consderable degradation in the phase-noise of the QVCO.

Regarding the LC VCO of Fig. 2.2 as an injection-locked oscillator (ILO), we note from Fig. 2.3 that the injection current in this ILO is 90° out of phase with the output voltage. From injection-locking theory [19], this implies that the ILO is operating at the edge of its lock range. Hence, the operating frequency of the QVCO is not exactly the same as the center frequency of the tank circuit [16]. Thus the effective

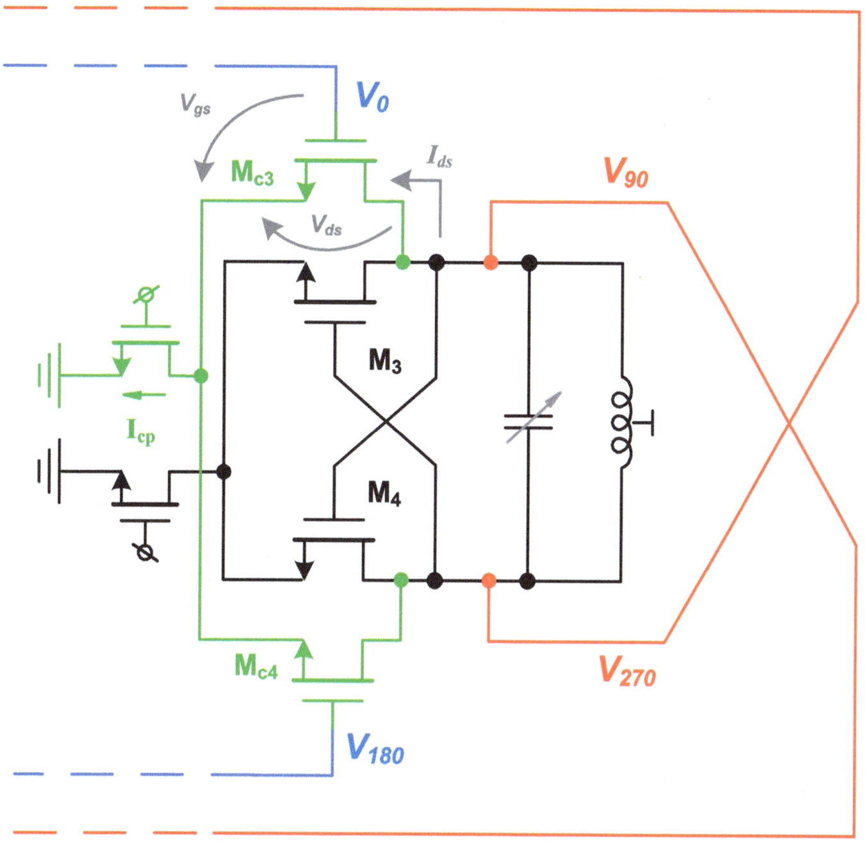

Fig. 2.2 Injection voltages and currents in one-half of a QVCO

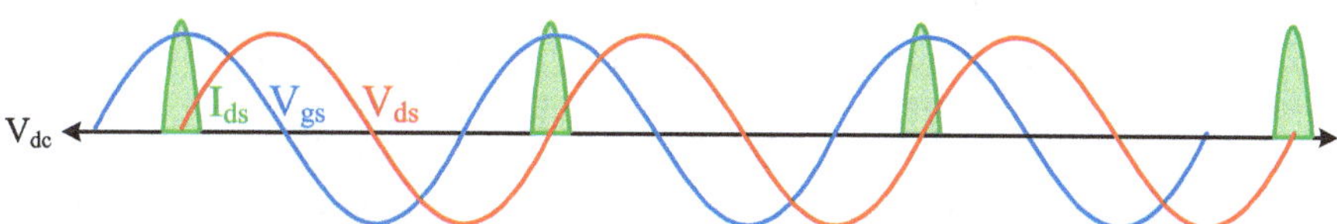

Fig. 2.3 Gate and drain voltages, and drain current of transistors M_{c3} and M_{c4}

tank Q at the QVCO's running frequency is lower than its peak value, leading to more degradation of phase-noise performance. Moreover, by changing the tail-current of the injection-pair, the lock range of each ILO changes resulting in a change in the QVCO frequency. This "varactor-like" effect leads to flicker-noise upconversion, adding $1/f^3$ phase-noise [20].

Several techniques were presented in literature to overcome the shortcomings of the basic LC QVCO. The rest of this chapter presents an overview, and classification of these techniques [21].

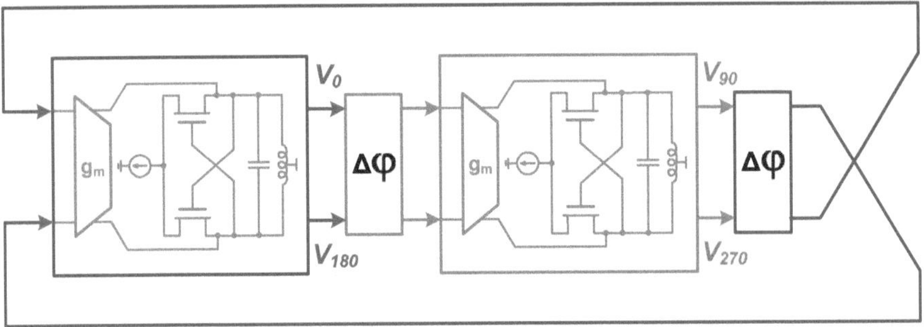

Fig. 2.4 Conceptual phase-shift based LC QVCO

2.2 Classification of QVCO Techniques

This section discusses the techniques used in literature to improve the performance of the basic LC QVCO [12, 20, 22–31]. These techniques can be classified into three broad categories: phase-shift based techniques, super-harmonic coupling based techniques, and direct-coupling-variant techniques. The following subsections give an overview of these techniques. Due to the large number of publications, the overview is not meant to be comprehensive.

2.2.1 Phase-Shifting Based QVCOs

The basic block diagram of this class of LC QVCOs is shown in Fig. 2.4. Similar to the basic LC QVCO, the phase-shift based LC QVCO consists of two injection-locked oscillators with both direct and cross-coupling. However, the output of each ILO is phase-shifted by $\Delta\varphi$ before injecting it into the next ILO. This means that the injection current in each tank *(is not)* in quadrature with the tank's output voltage. Rather, injection current is shifted from quadrature by an additional angle equal to the phase shift ($\Delta\varphi$) (this is equivalent to shifting I_{ds} in Fig. 2.3 by an additional $\Delta\varphi$). Ideally, a phase shift of 90° would make the injection current coincide with the peak of the output voltage, resulting in minimal phase-noise penalty [18]. Moreover, the 90° phase shift would cause the QVCO frequency to coincide with the tank frequency, maximizing the Q-factor and eliminating the "varactor-like" effect and its consequent $1/f^3$ phase-noise. A 90° is hard to achieve in practice, however. Nevertheless, a phase shift in the order of 40°–50° can help decrease the phase noise injected from one oscillator into the other [20].

One possible way to implement the phase-shift is through the use of R–C degeneration in the coupling transistors as shown in Fig. 2.5 [20]. The injection tail-current source is split in two to allow the use of R–C degeneration without hurting the headroom. Assuming the values of the resistor and the capacitor in Fig. 2.5 are $2R_s$ and $C_s/2$ respectively, then the resultant transconductance can be given by [20]:

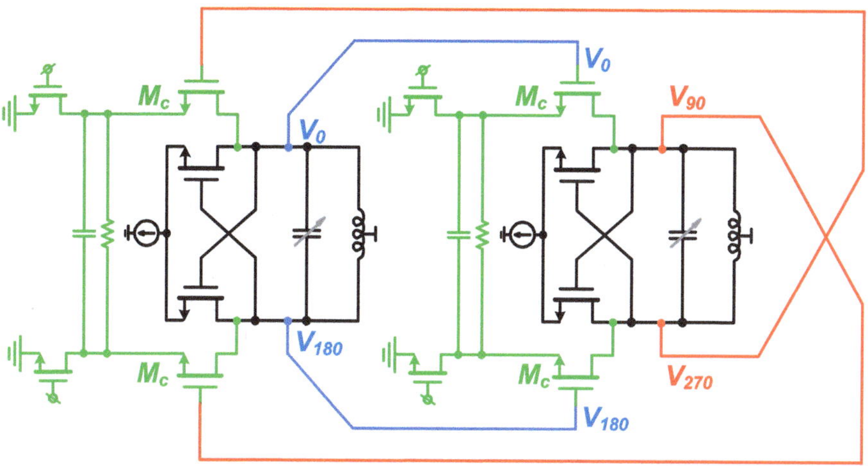

Fig. 2.5 Phase-shift based QVCO with R-C source-degeneration

$$G_m = \frac{g_m}{1 + g_m R_s} \cdot \frac{1 + sR_sC_s}{1 + \left(\frac{sR_sC_s}{1+g_mR_s}\right)} \tag{2.2}$$

where G_m is the equivalent small-signal transconductance of the coupling structure, and g_m is the small-signal transconductance of the coupling transistors. It is clear from Eq. 2.2 that G_m has a zero and a pole, making a 90° phase-shift impossible. With proper choice of R_s and C_s, however, a phase shift of 40°–50° is achievable [20]. A clear disadvantage of this architecture is evident from Eq. 2.2; the achieved phase-shift is frequency dependent, necessating a careful design of the phase-shift network.

An alternative implementation of a phase-shift based LC QVCO is shown in Fig. 2.6 [22]. Resistors are added in series with the gates of the coupling transistors, forming an R–C network with the parasitic C_{gs}. By adjusting the resistance value, the desired phase-shift can be achieved. The drawback, however, is that the Q of the parasitic C_{gs} capacitor is reduced due to the resistance. Since this parasitic capacitor is part of the tank, this means that the whole tank Q is reduced. Hence, to minimize this effect, the C_{gs} capacitance has to be much smaller than the tank capacitance [22]. The phase-shift is frequency dependent as well [20].

Several other phase-shift implementations are worth mentioning. In [23], phase-shift is implemented in two-steps. First, two quadrature currents are added generating a 45° phase-shift. Capacitive degeneration then *"ideally"* adds another 45° to get a total of 90° shift as shown in Fig. 2.7a. It is to be noted, however, that the phase-shift in this architecture is still frequency dependent (g_m/C is designed to be equal to the QVCO center frequency). Another drawback of this architecture is that eight transistors are used for coupling instead of four, resulting in more noise.

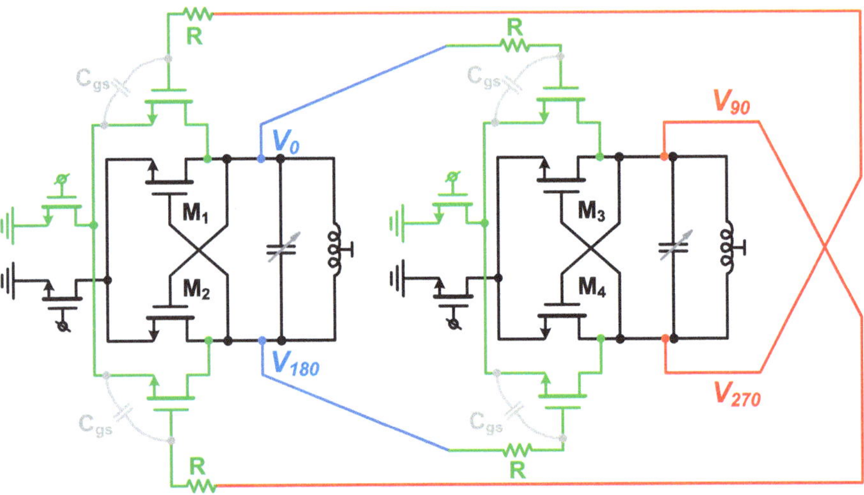

Fig. 2.6 Phase-shift based QVCO using resistors in coupling path

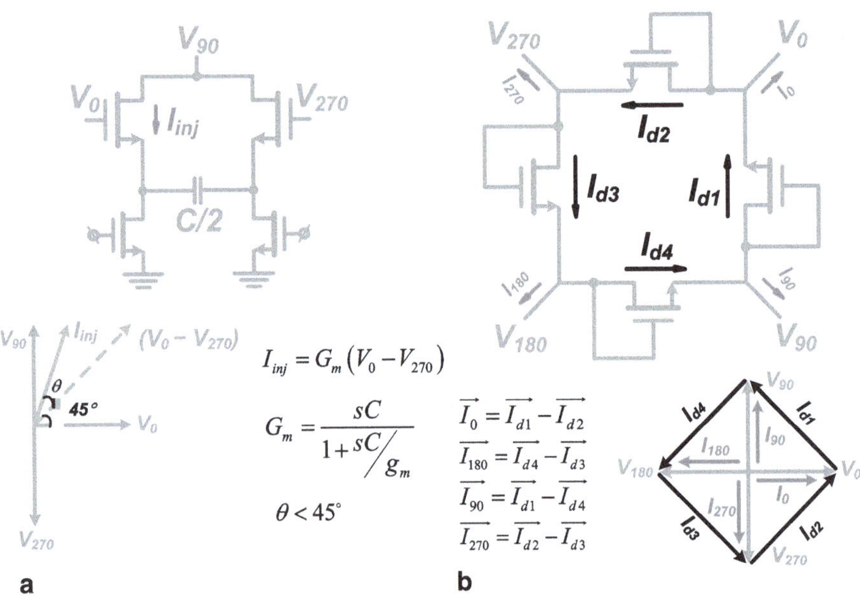

Fig. 2.7 Phase-shift QVCO with **a** current summation and capacitive degeneration and **b** frequency-independent current injection

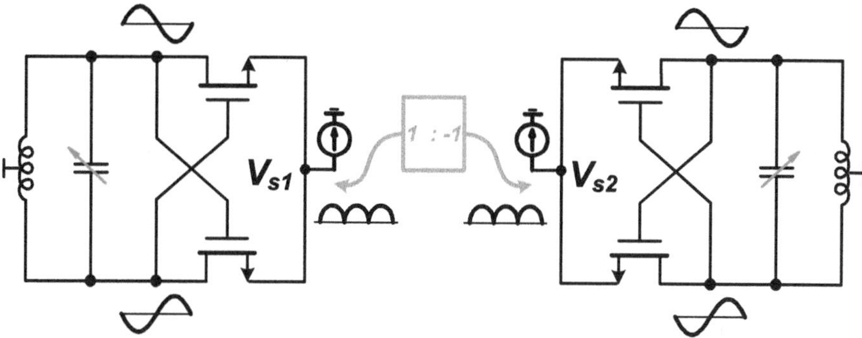

Fig. 2.8 Conceptual super-harmonic-coupling based LC QVCO

A frequency-independent phase-shifting architecture is presented in [24]. A symmetric coupling network is formed using only transistors in class-C operation as shown in Fig. 2.7b. The network is formed such that the injection current at each node is the sum of two currents: one leading the node's voltage by 45°, and the other lagging the node's voltage by 45°. Consequently, the resultant injection current at each node is in phase with the node voltage [24], as evident from the phasor diagram in Fig. 2.7b. Possible drawbacks are: the need of relatively large signal swing (to support class-C operation), as well tank de-Q'ing due to non-linear loading by diode-connected devices.

2.2.2 Super-Harmonic-Coupling Based QVCOs

A general representation of this class of LC QVCOs is shown in Fig. 2.8. Due to the hard-switching of the cross-coupled pair in the LC VCO, the common-source node is not a virtual ground as in a small-signal differential pair. During each half-cycle, one of the cross-coupling transistors is on and the other is off. The on transistor, together with the tail current source, forms a source follower that transfers the input voltage to the common-source node. This happens twice during each switching cycle, leading to an effective doubling of frequency at the common-source. The major premise of the super-harmonic coupling LC QVCO is to couple the common-source nodes of two LC VCOs in an anti-phase fashion. By forcing a 180° phase-shift between the common-source nodes (at double the output frequency), a 90° phase-shift is ensured between the outputs of the two LC VCOs.

One of the earliest implementations of the super-harmonic coupling concept is shown in Fig. 2.9 [25]. The tail current sources are removed and replaced by tail inductors. The tail inductors resonate with the parasitic capacitance at the common-source nodes at twice the QVCO's frequency. This forms a filter that improves the phase-noise of the oscillator. Moreover, removing the tail current source eliminates

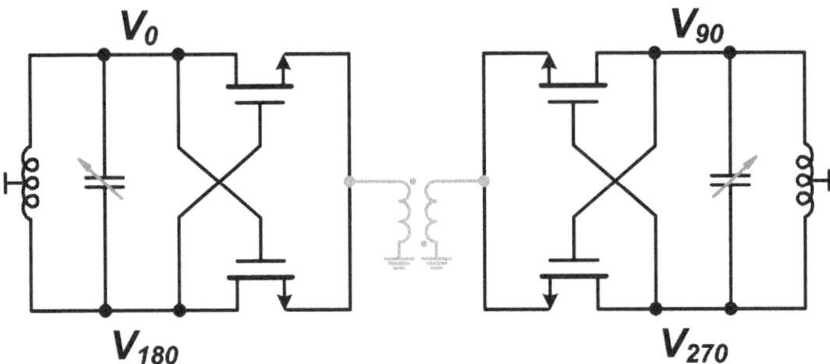

Fig. 2.9 Super-harmonic coupled QVCO employing transformers

one of the largest sources of flicker noise, effectively reducing the $1/f^3$ phase-noise [32]. By coupling the two tail inductors in an inverting transformer structure, the required anti-phase coupling is achieved. A simple, and area-efficient way to implement the transformer is through the use of a single symmetric inductor [25]. Unlike the basic LC QVCO, this coupling method does not inherently shift the oscillation frequency away from the tank center frequency. Hence, the maximum tank Q is utilized, leading to a good phase-noise performance. The drawback, however, is that lower oscillation amplitudes can result in even-mode operation leading to in-phase operation (instead of quadrature). While this can be mitigated by ensuring a large oscillation amplitude [25], it can put a limit on the minimum achievable frequency at constant power (amplitude decreases as frequency decreases for constant power in an LC VCO).

An alternative implementation of the super-harmonic coupling concept is shown in Fig. 2.10 [26]. It looks very similar to the basic LC QVCO of Fig. 2.1. However, it has three distinctive differences: the coupling transistors are PMOS as opposed to NMOS, the drains of the coupling transistors are tied together to the common-source nodes of each of the two LC VCOs, and the sources of the coupling transistors are tied to the supply. In fact, each of the coupling transistor pairs ($M_{c1} - M_{c2}$ and $M_{c3} - M_{c4}$) forms a frequency doubler. The outputs of each of the two LC VCOs are frequency-doubled, and the doubled output is fed to the common-sources of the two LC VCOs in anti-phase. This avoids the use of passive transformers, which saves area. Moreover, using PMOS transistors for coupling makes them nominally off (both gate and source are at supply voltage under no oscillation) leading to a class-C operation which reduces the power overhead associated with the use of active components. In addition, the cycling switching of the transistors reduces their inherent $1/f$ noise content [26]. Again, the drawback is the need for high oscillation amplitude for strong coupling. Moreover, the use of active devices, while reducing area, adds more noise. Besides, active devices add extra loading which reduces the achievable tuning range.

Figure 2.11 shows another implementation of the super-harmonic coupling concept [27]. The anti-phase coupling is ensured by cross-coupling of the tail current

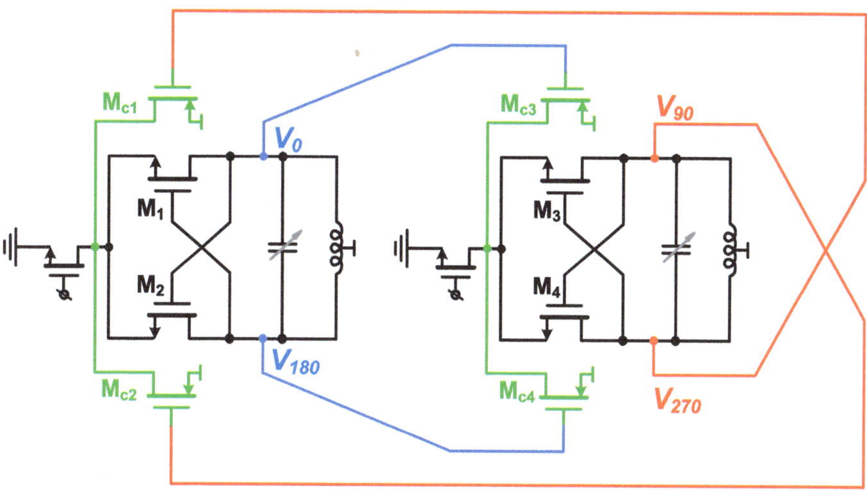

Fig. 2.10 Frequency doubler based super-harmonic QVCO

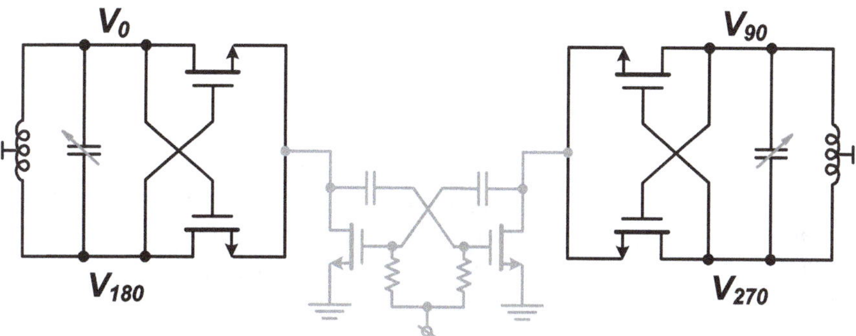

Fig. 2.11 Super-harmonic QVCO with cross-coupled tail transistors

sources. This ensures that the drains of the tail current-sources (which are the common-source nodes of the VCOs) are out of phase. Again, the swing needs to be high to ensure proper quadrature coupling [27]. Another drawback is that the cross-coupling structure places a large capacitive load at the common-source node. While this is beneficial in reducing the phase-noise due to the tail current source, it can cause an effective de-Q'ing of the tank [32].

2.2.3 Direct-Coupling Variants

In this category of QVCOs, direct coupling is performed using configurations different than that in Fig. 2.1. For instance [12] uses series transistors, instead of a parallel

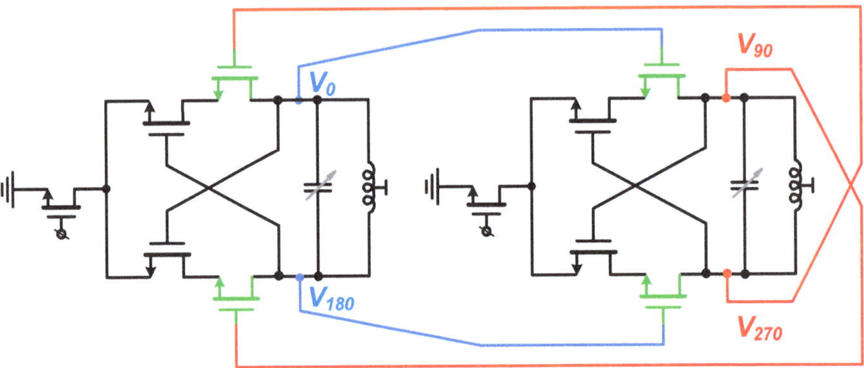

Fig. 2.12 Series-coupled LC QVCO (with coupling transistors on top)

differential pair, to perform direct coupling. As shown in Fig. 2.12, the coupling transistors are placed in series with, and on top of, the negative-g_m cross-coupled transistors. This configuration breaks the trade-off between phase-noise and quadrature accuracy, allowing relatively constant phase accuracy regardless of the coupling transistor sizes [12]. Besides, the series placement means that the coupling transistors can reuse the current in the negative-g_m transistors, leading to power saving. Nevertheless, for optimum phase-noise performance, relatively large coupling transistor sizes are needed (in the order of five times the size of the g_m transistors in [12]). The relatively large sizing (needed for stacking as well as phase-noise performance) adds large loading to the QVCO, limiting the maximum achievable tuning-range. Moreover, transistor stacking limits the available headroom and makes supply scaling difficult. Bottom series coupling (where coupling devices are placed below the negative-g_m devices) is also possible [28].

Another modified direct-coupling scheme is shown Fig. 2.13 [29]. This scheme exploits the fourth terminal (the bulk contact) of the g_m-cell transistor for coupling, eliminating the need for an additional coupling transistor. In addition to reducing the power consumption, this also removes the additional noise contributed by the coupling transistors. To allow this scheme in a conventional NMOS-based g_m-cell, the transistor has to be a triple-well transistor to allow for a separate bulk terminal (not tied to the substrate). Note that AC-coupling is used, where resistors are used to dc-bias the bulk terminal at its nominal ground potential. While triple-well devices are available in most of the RF technologies, they are not a standard option for a plain vanilla CMOS technology. Moreover, injecting the signal into the bulk incurs the risk of forward-biasing the bulk terminal, which is directly loading the tank. Hence, the Q of the tank can be reduced by the forward biased junction leading to a phase-noise degradation.

In [30], transformer coupling is used for injection instead of active devices as shown in Fig. 2.14. This is done by using transformer coupling between the sources of one stage, and the drains of the next stage. Hence, active coupling devices are eliminated together with their noise effect. Moreover, transformers do not limit the swing allowing for low supply voltages to be used. However, the operating frequency

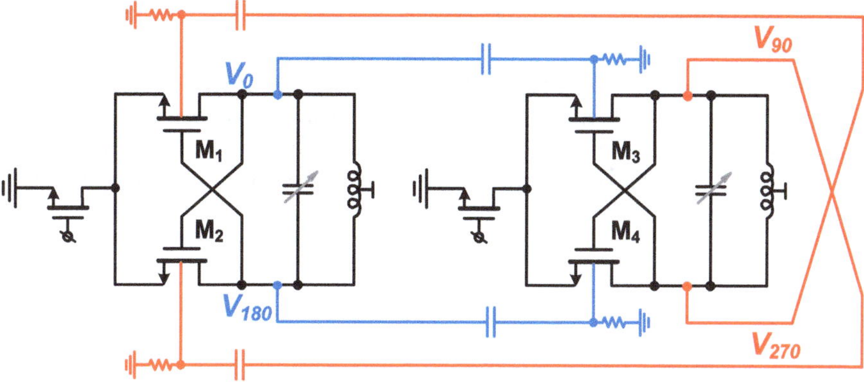

Fig. 2.13 LC QVCO with back-gate coupling scheme

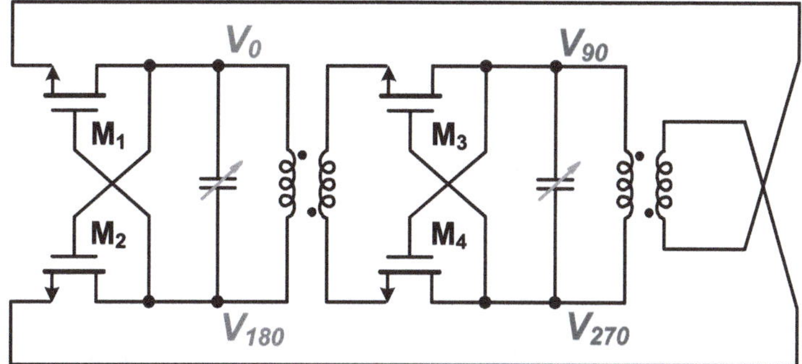

Fig. 2.14 Simplified schematic of transformer-coupled QVCO

does not coincide with the tank center frequency (similar to the basic QVCO). Hence, phase-noise is degraded due to this phase-shift effect [30]. This issue is addressed in [31], which uses a similar structure with two differences. First, no cross-coupling is used in the individual VCOs, forming a ring-oscillator structure instead of two coupled VCOs.[1] Second, each transformer forms a coupled resonator. With proper tuning of the coupling factor (k), the phase shift of the coupled resonator structure can be made to be 90° making the tank resonance and the QVCO frequency the same and hence improving the Q and the phase-noise performance. The drawback, however, is that relatively small coupling factors (in the order of 0.2 or less) are needed complicating the transformer design and requiring relatively tight control on the coupling factor value.

[1] A ring oscillator is a cascade of amplifiers whose gain is at least unity, and whose phase is zero, at the frequency of oscillation. The individual amplifiers do not oscillate on their own. In the other discussed QVCOs, the individual VCOs oscillate on their own (no cascade is needed) but are coupled through a network to generate quadrature phases.

2.3 Conclusion

In this chapter, the major drawbacks of the basic LC-QVCO were outlined. The different methods, used in literature to overcome these limitations, were also presented and classified. Based on the understanding developed in this chapter a new robust, method for quadrature VCO coupling, which lends itself to wideband opertion, is presented in the next chapter.

Chapter 3
QVCO with Complementary Coupling

3.1 Introduction

Software-defined radio allows a single front-end to be used for multiple standards, through software programmability and reconfigurability. The original SDR, proposed by "Mitola" [5], has an ADC directly following the antenna, allowing all downconversion and post-processing to be performed in digital domain. This approach, in theory, would provide the highest level of flexibility and programmability. However, no practical implementation is possible with current technology. A more practical approach is to use a generic direct-conversion architecture with programmable filters as well as programmable LO to cover multiple bands [6].

For very wideband programmability in the LO path, banks of VCOs are needed [6]. Quadrature VCOs allow the direct generation of quadrature LOs without requiring the VCO to operate at twice the desired operating frequency. A wide tuning range in the QVCO allows less number of QVCOs in the bank, thus reducing area requirements, complexity, and design time. This chapter presents a wide tuning range quadrature VCO design for SDR-type applications in the lower GHz range (5–10 GHz).

3.2 Proposed Architecture

3.2.1 Motivation

In this section, we propose a new architecture for LC QVCOs, based on the phase-shifting concept outlined in Sect. 2.2.1. The motivation is achieving a QVCO architecture that is:

1. simple and robust
2. achieves good phase-noise performance
3. does not limit the tuning range
4. operates with low-supply voltage (1 V or less)

M. Elbadry, R. Harjani, *Quadrature Frequency Generation for Wideband Wireless Applications*, Analog Circuits and Signal Processing, DOI 10.1007/978-3-319-13788-9_3

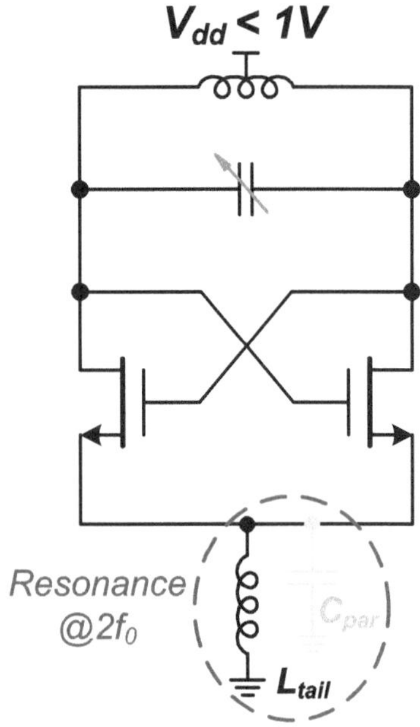

Fig. 3.1 Core VCO of the proposed scheme

The QVCO, in this work, is designed for use in a wideband phased-array receiver that covers the band from 6 to 8 GHz (instantaneous bandwidth is 2 GHz). Moreover, the technology in use is TSMC 65 nm CMOS technology with a nominal supply voltage of 1 V requiring the QVCO to operate at, or below, 1 V.

3.2.2 Core VCO Choice

The core VCO is a simple NMOS cross-coupled LC VCO, with no tail source as shown in Fig. 3.1. An inductor is placed at the common-source which forms a resonant circuit with the total parasitic capacitance at that node (C_{par}), at twice the oscillation frequency $2f_0$. In addition to allowing for a larger swing, removing the tail source also greatly reduces the $1/f^3$ noise [18, 32]. With large oscillation swings, the cross-coupled g_m transistors go into triode region for a considerable part of the cycle. The resonant circuit adds a high impedance in series with the low-resistance switch, avoiding Q factor degradation at high swing and, improving the overall phase-noise performance [32].

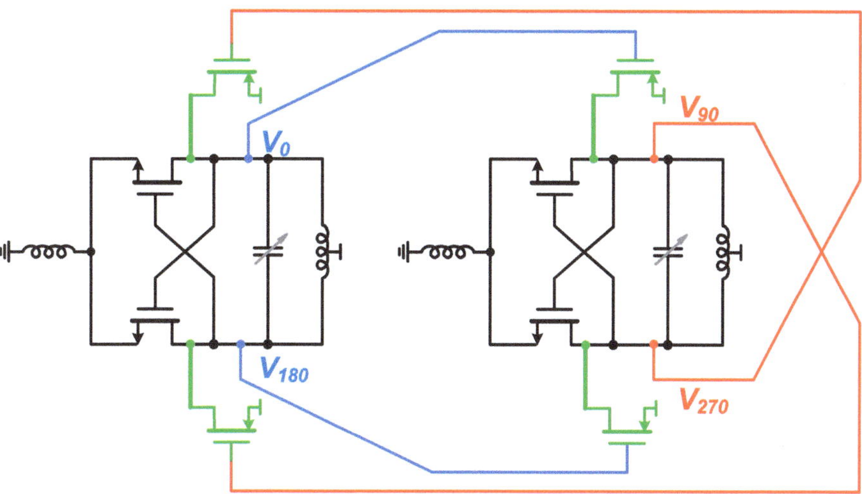

Fig. 3.2 The proposed QVCO

3.2.3 *Full QVCO*

The schematic of the proposed QVCO is shown in Fig. 3.2. The QVCO is formed by coupling two of the core VCOs in Fig. 3.1. The proposed QVCO resembles the direct-coupled QVCO of Fig. 2.1 with one difference; the coupling transistors are *PMOS instead of NMOS*. While this might seem to only shift the injection current by 180° making the proposed QVCO very similar to the basic QVCO, the direct intuition is not correct. The use of complementary devices for coupling the two core VCOs[1], results in an implicit phase-shifting effect that pushes the injection current away from the voltage zero-crossing points of the oscillator's output. Thus, this architecture can be classified under the "Phase-shift" category. Unlike the architectures presented in Sect. 2.2.1, however, the phase-shift is not performed by passive devices (which are frequency-sensitive), nor does it need a special coupling network that might de-Q the tank (as in [24]).

To understand the phase-shifting effect of the proposed QVCO , the large-signal voltage and current waveforms of the coupling transistors have to be studied. Towards this end, consider Fig. 3.3 which depicts the large-signal time-domain waveforms of the coupling transistors. Due to the use of PMOS (or more generally: complementary coupling), the coupling transistor is nominally off; both the source and gate voltages sit at the supply value in absence of oscillation setting V_{sg} to zero. Similarly, the source and drain voltages sit at the supply value making for a zero V_{sd}.

In the oscillation mode, the V_{sg} and V_{sd} voltages are in quadrature since they are the single-ended quadrature outputs of the oscillator. When the V_{sg} voltage of the coupling transistor is at its peak (where it should be the most conductive), the V_{sd} voltage is zero, thus forcing the I_{sd} current to zero. *Hence, injection current is forced to zero at the zero-crossings of the oscillator's output voltage.* Away from the zero crossings, current is injected when $V_{sg} > V_{th}$ and V_{sd} is non-zero (positive

[1] This means that if NMOS devices are used for the g_m-cell, PMOS devices are used for coupling and vice-versa.

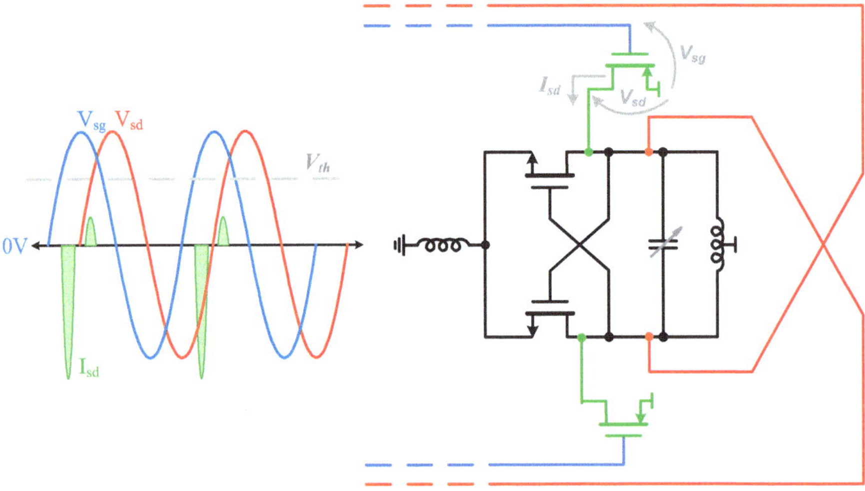

Fig. 3.3 Current and voltage waveforms of the coupling transistors

or negative). As shown in Fig. 3.3, this results in two current pulses during each oscillation cycle, with one pulse (corresponding to a higher $|V_{sd}|$) much higher than the other. Both current pulses are injected significantly far from the zero-crossing point.

This in effect creates a phase-shift that reduces phase-noise, without resorting to frequency-sensitive passives, or the use of coupling networks that might reduce the tank Q, allowing for a wide-tuning-range QVCO with robust active injection. As an added advantage, PMOS devices have inherently lower flicker noise than their NMOS counterparts due to the buried nature of the channel. The drawback, however, is that PMOS devices need to be larger than equivalent NMOS devices for equal injection strength. This becomes a problem only when the active devices constitute a large percentage of the total tank capacitance (e.g. for mm-Wave oscillators).

Comparison: Normal and Complimentary Coupling

To evaluate the effectiveness of the proposed scheme, simulations are made to compare the performance of the proposed QVCO (of Fig. 3.2), with a similar QVCO that has the PMOS coupling transistors (green transistors in Fig. 3.2) replaced with NMOS transistors. NMOS transistors have higher mobility than PMOS transistors. Hence, to ensure a fair comparison, the NMOS transistors are sized so that their total injected charge (integration of the injected current over time) is equal to their PMOS counterparts. Both oscillators are operated at an 8 GHz frequency.

Figure 3.4 shows the injection current waveform overlapped with the output voltage waveform for PMOS coupling. The trends suggested in Fig. 3.3 can clearly be observed; the injection current is very small at the voltage zero-crossings, and is

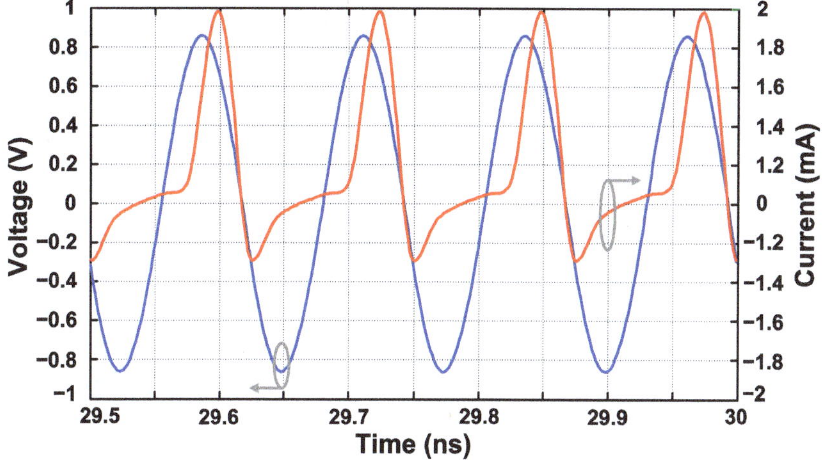

Fig. 3.4 Injection current and output voltage for PMOS coupling (simulated)

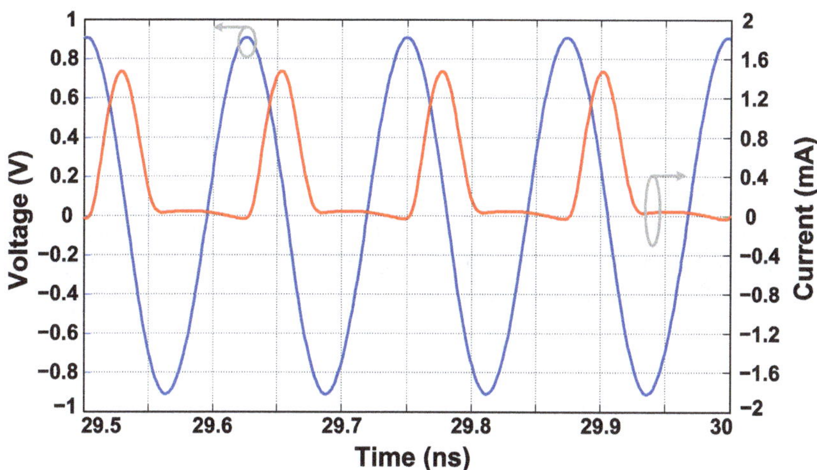

Fig. 3.5 Injection current and output voltage for NMOS coupling (simulated)

maximum away from the crossing. The peak current is shifted from the voltage zero-crossing by around 55° (with no explicit phase-shifting network employed). Similar waveforms for the NMOS case are shown in Fig. 3.5, showing that the injection current peak almost coincides with the output voltage zero-crossing. The peaks are not perfectly aligned; there is a finite phase-shift of around 10°. Nevertheless, PMOS coupling adds significantly higher shift, moving the injection current peaks closer to the voltage peaks.

To assess the impact of the phase-shifting effect on the phase-noise performance, simulations are done for both PMOS and NMOS coupling cases. As mentioned

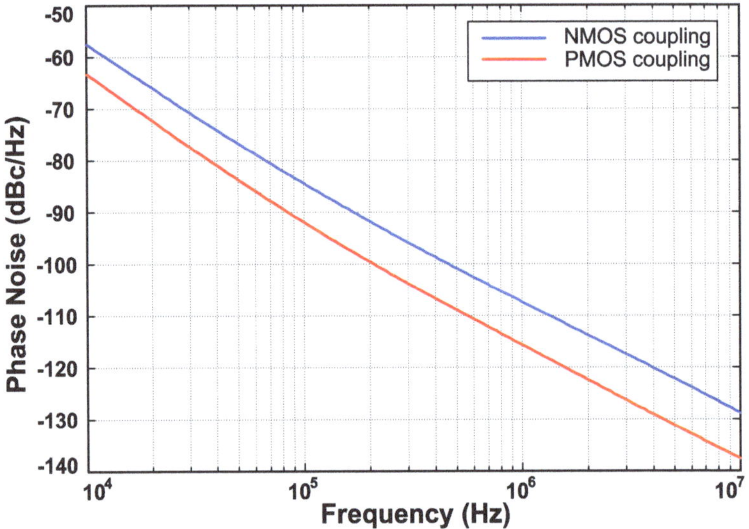

Fig. 3.6 PMOS and NMOS coupled QVCOs phase-noise

before, both NMOS and PMOS coupled QVCOs are sized to have equal injected charge, and are tuned to oscillate at the same frequency of 8 GHz. As evident from Fig. 3.6, the PMOS coupled QVCO has significantly better phase noise performance as compared to its NMOS counterpart over three decades of frequency offset. The PMOS coupled version is better by 6 dB in the $1/f^3$ region, and the improvement goes up to 8 dB in the $1/f^2$ region. Clearly, PMOS coupling is advantageous to NMOS coupling in terms of phase-noise.

3.3 Prototype Design

Based on the proposed complementary coupling architecture, a prototype QVCO is designed in TSMC 65 nm CMOS technology. Targeted at a wideband phased-array application, the design aims at achieving a tuning range of 6–10 GHz (to cover the band from 6 to 8 GHz with margin for process shifts). Injection locking is used in the phased-array receiver, hence no varactor tuning is needed. To increase the injection current, the PMOS coupling transistors are AC coupled and DC-biased at mid-supply so that their quiescent current is non-zero as shown in Fig. 3.7. Hence, the PMOS coupling transistors operate in a class-AB mode instead of class-C operation.

A 4-bit binary-weighted MIM-capacitor bank is used for tuning. To ensure operation at the intermediate point between current-limited and voltage-limited regimes (for best phase-noise performance [33]), the QVCO operates at a supply voltage of 0.55 V. The capacitor bank switches, however, are operated at the full supply voltage

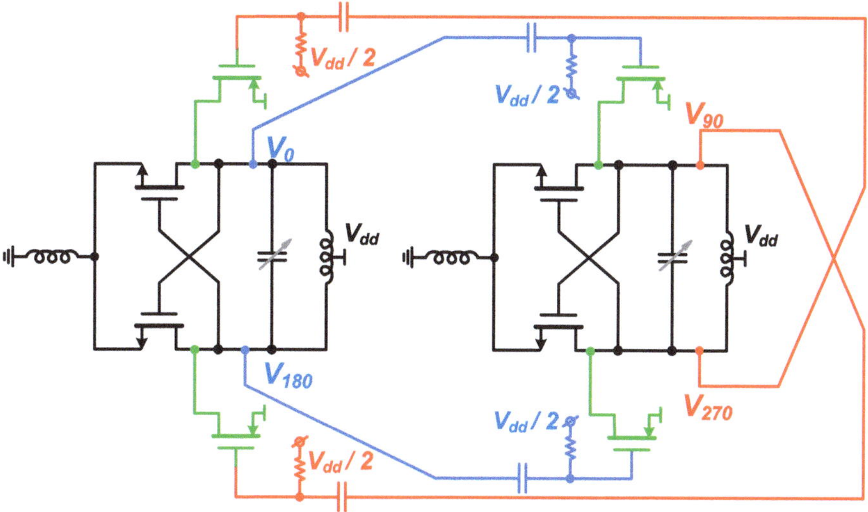

Fig. 3.7 Designed prototype with class-AB complementary coupling

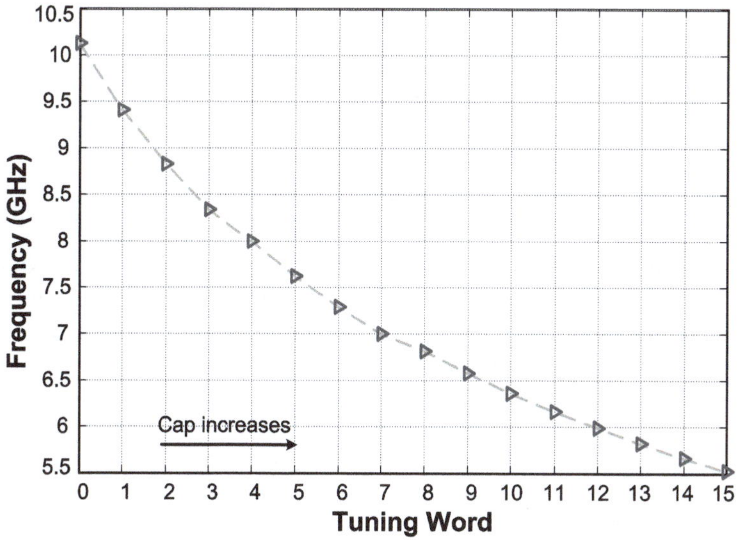

Fig. 3.8 Output frequency of the prototype versus tuning word

of 1 V to ensure the lowest possible on-resistance for a given switch size, and hence the highest over-all tank Q.

Extracted simulation results of the discrete tuning characteristics of the designed prototype are shown in Fig. 3.8. The QVCO can be tuned from 5.5 GHz at the lowest frequency end (larger tuning word and capacitance value) to 10.1 GHz at the highest

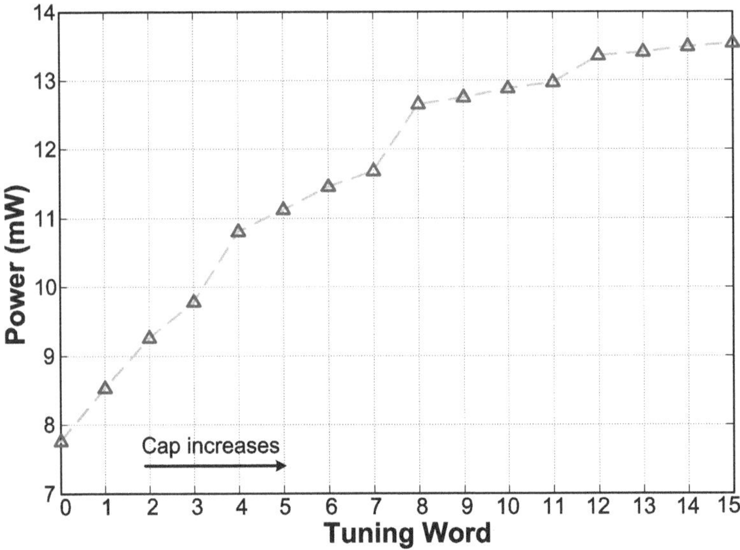

Fig. 3.9 Power consumption of the prototype versus tuning word

frequency end (where the load capacitance is purely from parasitics). Hence, the proposed VCO achieves a frequency tuning range (FTR) of approximately 59 %, where FTR is defined as:

$$FTR = \frac{2\,(f_{\max} - f_{\min})}{f_{\max} + f_{\min}} \tag{3.1}$$

The power consumption of the proposed QVCO versus tuning word is shown in Fig. 3.9. As the load capacitance increases (i.e. as frequency decreases), the power consumption increases. This can be explained as follows: with an almost constant tank-Q, the tank's equivalent parallel resistance R_p can be given by:

$$R_p = Q\omega L = Q\sqrt{\frac{L}{C}} \tag{3.2}$$

Since power consumption is inversely proportional to R_p (larger R_p requires smaller drive current for the same swing and vice-versa), this makes power consumption proportional to $\sqrt(C)$ (assuming a fixed voltage swing). In our case, swing also varies with frequency so the $\sqrt(C)$ dependence is not strictly valid. The trend, however, of higher power consumption for higher capacitance is still valid.

The simulated phase-noise of the proposed QVCO is shown in Fig. 3.10, at two offsets: 1 MHz and 10 MHz (post-layout). To get the best phase-noise performance at each discrete tuning-frequency, the supply voltage has to be changed to ensure that the QVCO is at the transition point between current-limited and voltage-limited regimes at that frequency. This was not done in simulation, however, to keep the

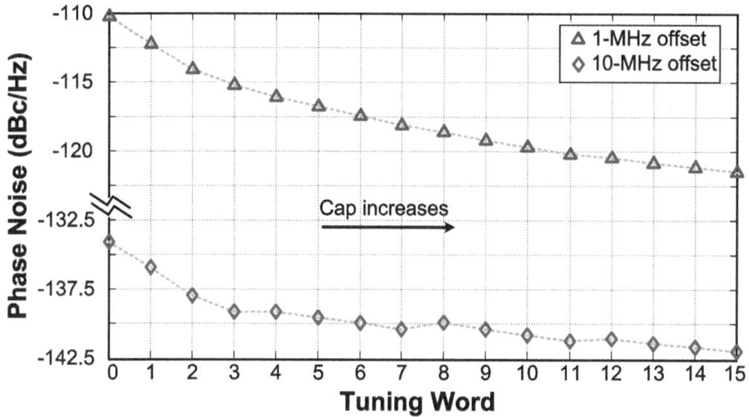

Fig. 3.10 Phase-noise of the QVCO prototype versus tuning word

results close to an actual practical scenario. Nevertheless, very good phase-noise performance is observed.

Since oscillator design entails a fundamental trade-off between frequency of operation, phase-noise, and power consumption, it is important to consider all these factors when evaluating the performance of an oscillator. A popular figure-of-merit (FOM) that takes these trade-offs into account can be given by [32]:

$$FOM = -\mathcal{L}(f_m) + 20Log\left(f_o/f_m\right) - 10.Log\left(\frac{P_{dc}}{1mW}\right) \tag{3.3}$$

where f_o is the VCO's center frequency, f_m is the offset frequency at which phase-noise is measured, P_{dc} is the power consumption, and $\mathcal{L}(f_m)$ is the phase-noise at offset f_m in dBc/Hz. This figure of merit does not capture an important design parameter: the VCOs tuning range. To capture the tuning range, another figure of merit is defined: the tuning-range figure-of-merit (FOMT), which can be given by [34]:

$$FOMT = FOM + 20.Log\left(\frac{FTR}{10}\right) \tag{3.4}$$

The FOM, for 1 MHz and 10 MHz offset frequencies, is shown in Fig. 3.11. The FOM at 1 MHz offset has a minimum value of 181 dBc/Hz, and decreases at higher frequencies (lower tuning word). This is attributed to the use of a constant supply to emulate a practical working environment where it would be difficult/impractical to tune the supply voltage for each frequency point. With a wide FTR of 59 %, the FOMT is 196.4 dBc/Hz and 200.4 dBc/Hz (worst-case) at 1 and 10 MHz offsets, respectively. This is at par with state-of-the-art validating the usefulness of the proposed complementary coupling technique.

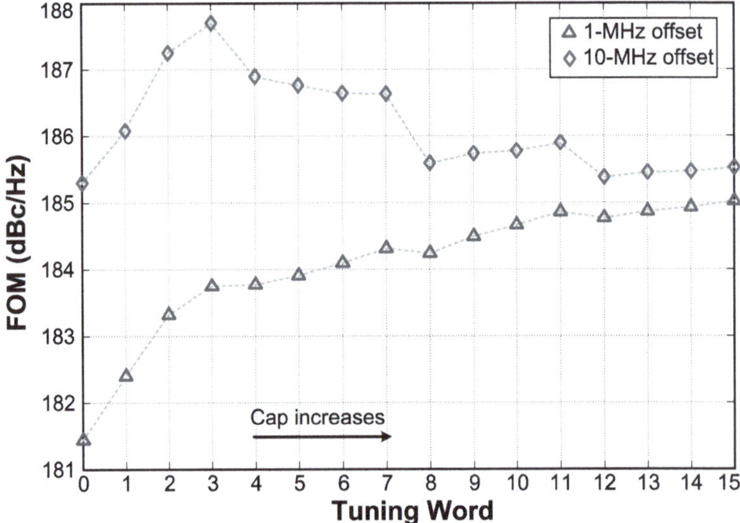

Fig. 3.11 Figure Of Merit (FOM) of the QVCO prototype versus tuning word

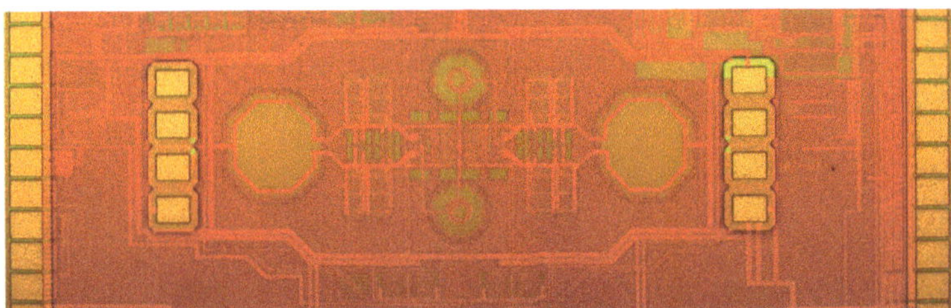

Fig. 3.12 Complementary QVCO chip micrograph

3.3.1 Measurement Results

The prototype was fabricated in TSMC 65 nm CMOS process. The design occupies an active area of 0.67 mm^2. The buffered QVCO outputs are connected to GSSG pads for on-chip probing, using 50 Ω buffers. The chip micrograph is shown in Fig. 3.12.

The fabricated QVCO can operate down to a 0.42 V supply. The QVCO is tested at three different supply voltages: 0.42 V, 0.5 V, and 0.6 V. In all cases, however, the supply voltage used for the switches in the capacitor array is kept at 1 V. The bulk of the PMOS couping transistors is also tied to 1 V. The supply voltage is varied by means of an off-chip linear regulator. The output of the GSSG probe is connected to an off-chip balun for differential to single-ended conversion and the single-ended output is fed to a R&S FSW43 spectrum analyzer for frequency and phase-noise measurement. The output frequency versus tuning word for the QVCO is shown in Fig. 3.13 for the 0.5 V supply.

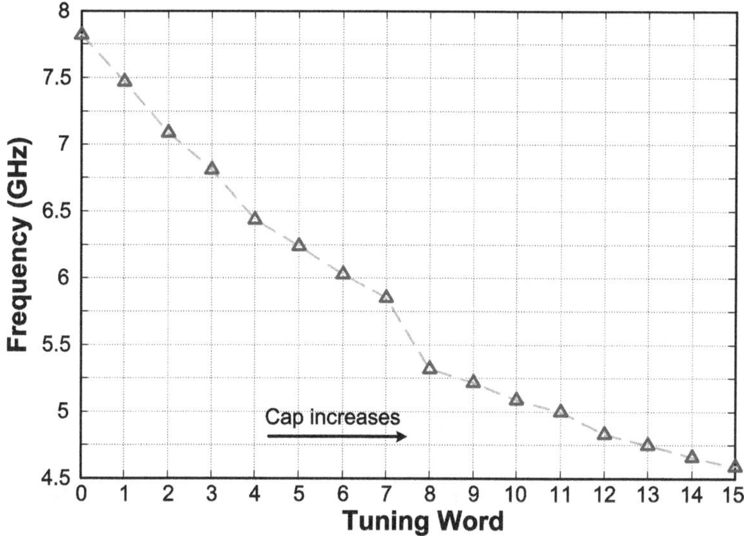

Fig. 3.13 Measured oscillation frequency at 0.5 V supply

Due to a problem with EM extraction, extra inductive parasitics were not accounted for leading to a frequency shift from post-layout simulation values. Nevertheless, the QVCO can be tuned from 4.59 to 7.82 GHz achieving an FTR of 52 %. Inspite of the shift, the tuning range is large enough to allow proper operation of the QVCO in the overall system. Similar results are obtained for the 0.42 V, and the 0.6 V, supplies. The output frequency for 0.42 V supply ranges from 4.57 GHz at the lowest end to 7.84 GHz at the highest end, achieving an FTR of 52.7 %. For the 0.6V supply, the tuning range is from 4.61 to 7.89 GHz with an FTR of 52.5 %.

The measured power consumption of the QVCO versus the tuning word is shown in Fig. 3.14 for the 0.5 V supply. Power consumption decreases for higher output frequencies (lower tank capacitance), and vice versa. The power consumption ranges from 7.36 mW at the highest frequency, to 10.98 mW at the lowest. Power consumed at 0.42 V supply ranges from 4.56 to 6.6 mW, whereas power consumed at 0.6 V supply ranges from 16.46 mW to 23.04 mW. To illustrate the relative magnitudes of the power consumptions at the different supplies, Fig. 3.15 shows the power consumption for the three different supply voltages on the same plot.

The measured phase-noise of the QVCO at 0.5 V, and tuning word equal to “5” (corresponding to 6.24 GHz, which is the mid-point of the tuning-range), is shown in Fig. 3.16. The frequency offset is swept from 300 to 100 MHz.

The phase-noise at 1 MHz and 3 MHz offsets, across the tuning range, is shown in Fig. 3.17 for 0.5 V supply. At 3 MHz offset, phase-noise ranges from −123.5 to −128.5 dBc/Hz across the tuning-range, while the phase-noise at 1 MHz offset ranges from −111.5 to −117.8 dBc/Hz across the tuning-range. For the 0.42 V supply, phase-noise at 3 MHz offset ranges from −121.7 to −127.3 dBc/Hz across

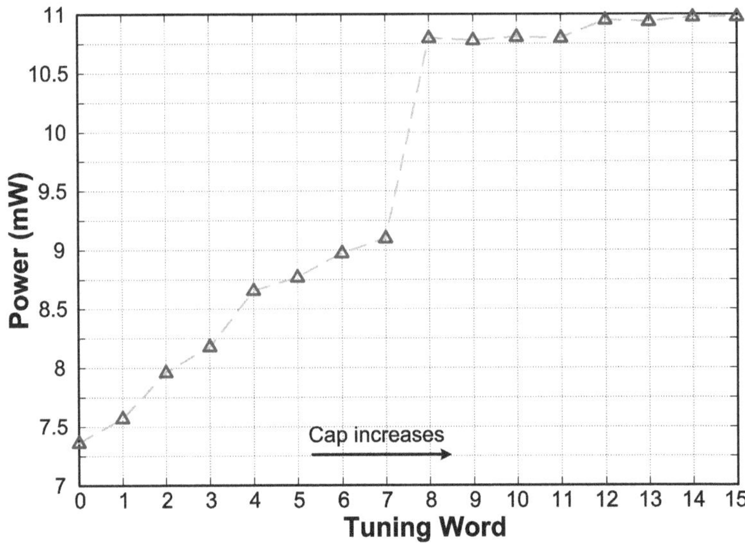

Fig. 3.14 Measured power consumption at 0.5 V supply

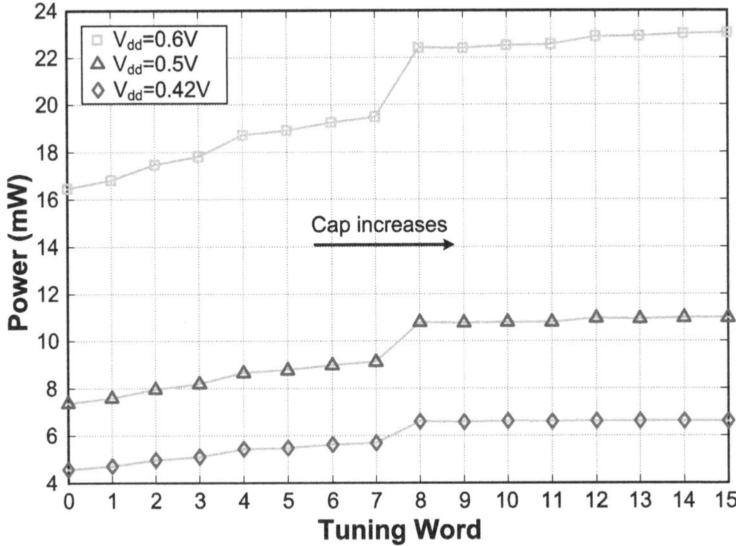

Fig. 3.15 Measured power consumption at 0.42 V, 0.5 V and 0.6 V supplies

tuning range whereas the phase-noise at 1 MHz offset ranges from −110.5 dBc/Hz to −115.87 dBc/Hz across tuning-range. On the other hand, the 0.6 V supply translates to a phase-noise of −123.26 to −130.26 dBc/Hz, and −110.5 dBc/Hz to −119.8 dBc/Hz, at 3 MHz and 1 MHz offsets, respectively.

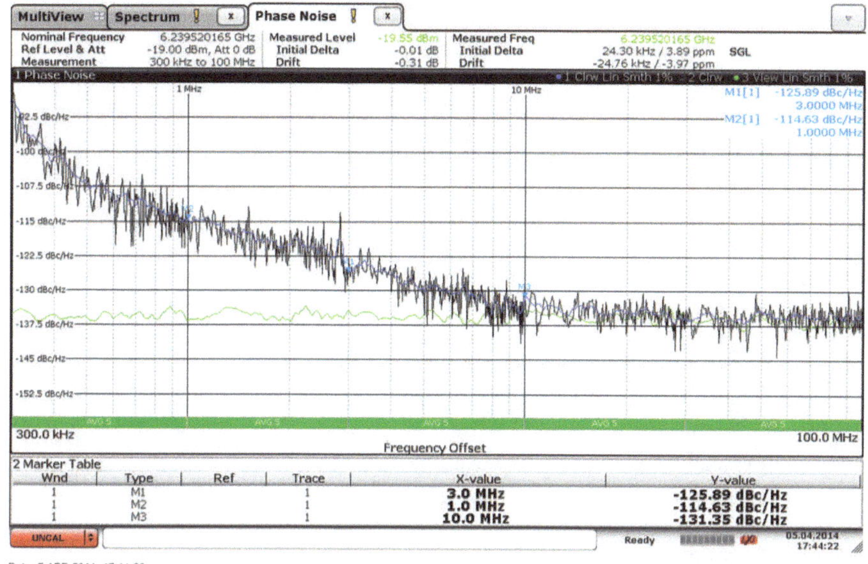

Fig. 3.16 Measured phase-noise at center of tuning-range with 0.5 V supply (tuning-word="5")

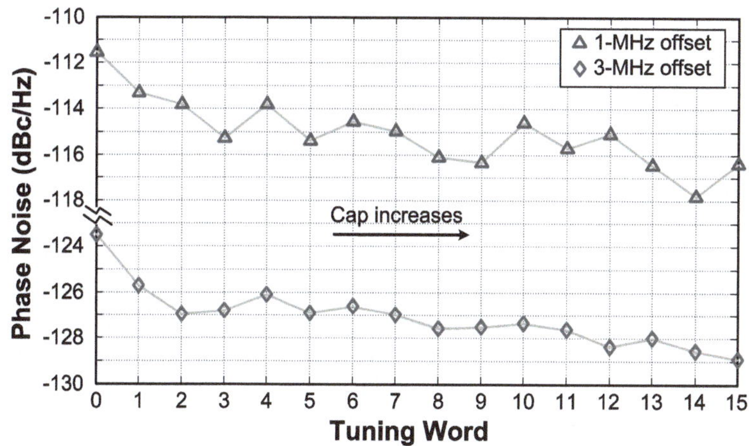

Fig. 3.17 Measured phase-noise at 1 MHz and 3 MHz offsetsat for 0.5 V supply

The FOM at 0.5 V supply (for 3 MHz offset) is shown in Fig. 3.18. The FOM touches 185.4 dBc/Hz at its peak, and goes down to 181.6 dBc/Hz at its minimum. The average FOM across the tuning range is 182.9 dBc/Hz. With a 52 % FTR, the average FOMT is 197.3 dBc/Hz (the peak is 199.7 dBc/Hz). Similarly, for 0.42 V supply the average FOM is 182.7 dBc/Hz with a maximum and minimum values of 184.7 dBc/Hz and 180.3 dBc/Hz respectively. The average, and peak, FOMT is

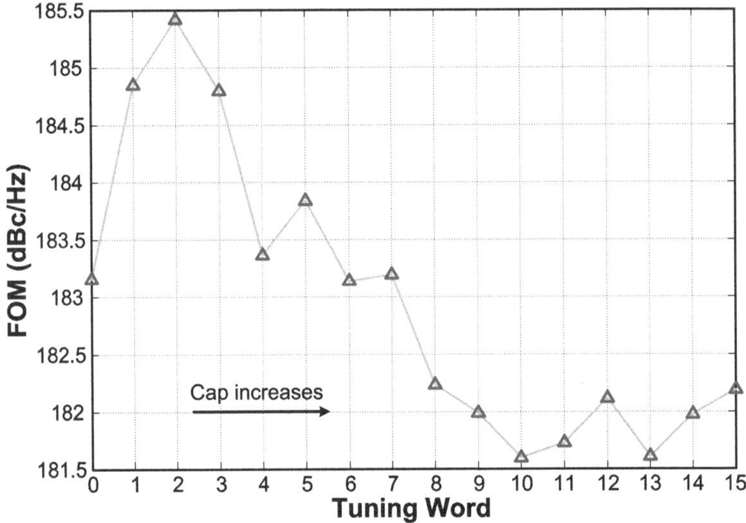

Fig. 3.18 Measured FOM at 3 MHz offset for 0.5 V supply

Table 3.1 Measured QVCO performance summary

Supply (V)	FTR (%)	Power (mW)	PN (dBc/Hz)	FOM (dBc/Hz)	FOMT (dBc/Hz)
0.42	52.7	5.5	−124.3	182.7	197.2
0.5	52	8.8	−126.9	182.9	197.3
0.6	52.5	17.8	−126.6	180.1	194.5

197.2 dBc/Hz and 199.2 dBc/Hz respectively. At 0.6 V supply, the average FOM degrades to 180.1 dBc/Hz with a maximum and minimum value of 181.6d Bc/Hz and 179.2 dBc/Hz respectively. The average, and peak, FOMT is 194.5 dBc/Hz and 196 dBc/Hz resepctively. The degradation at higher supplies comes from an increased power consumption without an equivalent phase-noise improvement.

The key performance aspects of the QVCO, at the three different supply voltages, are summarized in Table. 3.1. The power consumption and phase-noise numbers are those obtained at the mid-point of the tuning curve (a tuning-word of "5"). Phase-noise values are those obtained at 3 MHz offset. The FOM and FOMT values are the average across the tuning range. Optimal FOM can be observed both at the 0.42 V and the 0.5 V supplies. The 0.5 V supply, however, has the advantage of a lower phase-noise—without FOM degradation (i.e. extra power is justified for the phase-noise advantage). At 0.6 V supply, however, phase-noise performance is similar, whereas power consumption increases drastically, resulting in a lower FOM. With all the three supply voltages achieving almost the same FTR, the FOMT of the 0.6 V supply is also lower. Hence, the design-value 0.5 V supply is the optimum performance point in terms of achieving the best phase-noise performance without excessive power consumption, while covering the desired tuning range.

Table 3.2 Comparison with similar work

Ref	V_{dd} (V)	f_o (GHz)	FTR (%)	P_{dc} (mW)	PN (dBc/Hz)	FOM	FOMT
[36]	1.7	20.9	3.1	6.3	−126.5	195.6	185.4
[23]	2	1.57	24	30	−147.5	187.1	194.7
[25]	2.5	4.9	12.2	22	−134.5	185	186.7
[12]	2	1.8	18.3	25	−140	181.6	186.8
[29]	1.8	1.1	28	5.4	−137	181	190
[30]	1	17	16.5	5	−119.5	187.6	192
[35]	1.2	4.8	67	6–20	−123.6	176.5	193
This Work	0.5	6.25	52	7.4–11	−126.6	181.6–185.4	196–199.7

3.4 Conclusions

In this chapter, a novel complementary coupling technique for QVCOs was presented. Using a modification of the conventional active coupling, robust quadrature coupling together and low-phase noise are simultaneously achieved. The proposed technique, in essence, can be classified under the phase-shifting category. Nevertheless, no passive phase-shifting networks are used, hence the phase-noise advantage of phase-shifted coupling is achieved without the usual drawback of frequency sensitivity, and reduced tuning range. Furthermore, the architecture proposed is well-suited for low-voltage operation.

The performance of the fabricated prototype is compared with other QVCOs in published literature, in Table 3.2. The phase-noise values reported are those at 3 MHz offsets. If data at 3 MHz offset was not available, it was extrapolated with a 20 dB/decade profile. The supply used for the core of the QVCO in this work is 0.5 V, with the switched capacitor tuning useing a 1 V supply.

This work achieves the highest FOMT amongst similar work. It also has the second highest tuning-range, next to [35]. It is to be noted, however, that the large tuning range in [30] is achieved through the use of transformers; the tuning range in this work is achieved through the use of simple conventional tanks with no specially designed transformers. Moreover, the average FOM, as well as FOMT, of this design is higher. Furthermore, the achieved FOM is close to similar designs. Designs in [23, 25, 36] achieve higher FOM, but also use much higher supplies. The design in [30] achieves a higher FOM with the small supply voltage of 1 V but, again, at the cost of the added design complexity of using special coupling transformers.

In summary, the proposed technique provides state-of-the-art FOM performance with robust, simple design that uses conventional tank-circuits. Moreover, it achieves state-of-the-art tuning-range, with best-in-class FOMT performance.

Chapter 4
mm-Wave QVCO

4.1 Introduction

With the increasing demand for high data rates in wireless consumer products, there is a higher demand for wider RF bandwidth. Millimeter-wave bands meet these demands by providing wide RF bandwidths, allowing for Gb/s data rates. With 9-GHz of ISM (unlicensed) bandwidth between 57 GHz and 66 GHz, new standards have emerged to exploit the available bandwidth for high data rate applications leading to the development of the IEEE 802.11ad standard [37][1]; a WLAN standard complementing the, currently popular, 802.11a/b/g/n for short range, very high data rate (up to 6.75 Gb/s) applications. For wireless back-haul applications, the E-band spanning 71–76 GHz and 81–86 GHz allows 1Gb/s transmission over a distance of 2–3 km [38]. The millimeter wave band is also well-poised for radar applications. Two bands for vehicular radar are allowed by regulatory agencies: the 24 GHz and the 77 GHz bands [39]. The 77 GHz band[2] offers higher radar performance due to smaller antenna sizes, better angular resolution, smaller fractional bandwidth (hence easier passive component integration), as well as higher allowable transmit power (which translates to a longer range operation) [39]. CMOS technology has proven its viability for millimeter wave applications [40–43], allowing for high degree of integration and cost reduction of millimeter wave systems.

4.2 Motivation

Conventional quadrature generation techniques, polyphase filtering and divide-by-2 frequency dividers, are impractical for millimeter wave applications. Divide-by-2 quadrature generation requires the system's oscillator to operate at double the desired

[1] In US, only 7 GHz of bandwidth are available spanning 57 GHz–64 GHz.

[2] This band spans 76–77 GHz in US, and 77–81 GHz in Europe.

M. Elbadry, R. Harjani, *Quadrature Frequency Generation for Wideband Wireless Applications*, Analog Circuits and Signal Processing, DOI 10.1007/978-3-319-13788-9_4

output frequency, which is impractical or impossible for 60 GHz (and above) frequencies. Polyphase filtering requires the use of very small resistance and capacitance values, making them very sensitive to parasitics and mismatch effects. Quadrature generation, however, can be done by using 90° hybrids [40, 44], or quadrature VCOs [41, 42, 45]. Due to the passive nature of hybrids, however, their output suffers attenuation [44]. Quadrature VCOs are, thus, well-suited to millimeter-wave requiring less power for signal generation and buffering.

Several millimeter-wave QVCO designs are present in literature. In [46], the basic LC QVCO structure of Fig. 2.1 is used. A major drawback is the use of active coupling transistors that load the tanks, leading to a reduced tuning range. The "varactor-like" effect is used for additional tuning, by varying the tail current of the coupling transistors. While this adds a degree of freedom in tuning the oscillator, it comes at the cost of either a larger quadrature error (if the tail current is reduced to reduce the frequency) or a higher phase-noise (if the tail current is increased to increase the frequency). Thus, the optimal trade-off point between phase-noise and quadrature accuracy cannot be maintained across the whole tuning range. The QVCO has a center frequency of 48 GHz with a tuning range of 8 GHz achieving an FTR of 16.67 %.

In [47], the basic LC QVCO structure is also used. To cover the 57–66 GHz communications band, however, two QVCOs with overlapping tuning ranges are used. This incurs an area penalty, due to the replication of the QVCO as well as the VCO buffers. Nevertheless, a single QVCO just covers the desired 57–66 GHz band. The dual QVCO structure, however, allows an overall tuning range of roughly 57–72 GHz. Furthermore, the phase-noise performance is relatively poor which is characteristic of the conventional QVCO as discussed in Sect. 2.1.

In [31], a ring structure is used with transformer-based inter-stage coupling. The coupling factor is designed to induce a 90° phase-shift in-between stages to improve phase-noise as explained in Sect. 2.2.1. Although it achieves a very good phase-noise performance, the QVCO has a limited tuning range of 4.35 GHz falling short of covering the 57–66 GHz band. Besides, the transformers require a low-coupling factor to achieve the desired phase-shift, resulting in relatively complicated transformer design as well as signal loss.

In [48], the conventional QVCO structure is used along with transformer-coupling. Tuning is done through changing the tail current. The bi-modal operation of a QVCO is exploited to increase the tuning range. This is done by adding a switchable stage of 90° phase-shift. When the phase-shift stage is off, the QVCO operates in the "normal" mode corresponding to a center frequency higher than the tank's resonance frequency [15]. A major drawback is the dependence on a frequency-sensitive LC structure for the bi-modal operation, making the QVCO prone to mismatches and process variations (with no tracking between the main tank's LC and the phase-shift network's LC). Nevertheless, the QVCO achieves a tuning range of 13 GHz, covering the frequency range of 49–62 GHz with good phase noise performance.

In [24], the frequency independent phase-shift network shown in Fig. 2.7b is used to achieve good phase-noise performance over a wide tuning range. The QVCO can be tuned from 58 to 68 GHz. While the QVCO achieves a good phase-noise performance with relatively low power, the tuning range barely covers the desired 57–66 GHz band leaving little room for process shifts.

This work aims at achieving a wide-tuning range quadrature VCO that is well-suited to millimeter-wave operation, achieves reasonable phase noise and power numbers and, simultaneously achieving a tuning range that covers the 57–66 GHz communications band with enough room for process shifts [21]. Moreover, it is also desirable to achieve a tuning range that would allow the QVCO to be used in multiple millimeter-wave bands such as the 57–66 GHz ISM band, the 71–76 GHz E-band and the 76–77 GHz vehicular radar band.

4.3 Proposed QVCO

In the absence of parasitics, the ratio between the maximum and the minimum oscillation frequency of an LC oscillator would be given by:

$$\frac{f_{\max}}{f_{\min}} = \sqrt{\frac{C_{\max}}{C_{\min}}} \tag{4.1}$$

where f_{max} is the maximum oscillation frequency, f_{min} is the minimum oscillation frequency, C_{max} is the maximum capacitance, and C_{min} is the minimum capacitance. Practically, the tank capacitance has a considerable percentage of parasitics (C_{par}). If switched-capacitor tuning is used, then a part of the parasitic capacitance (C_{fixed}) will be due to cross-coupled pair, the parasitic capacitance of the tank inductor, input capacitance of the QVCO buffers, as well as any deliberate fixed capacitance connected to the tank. Another major parasitic component will arise from the parasitic top and bottom plate capacitances associated with the switched-capacitor bank, as well as the switch parasitics. If the parasitic capacitance for each switched capacitor of value C is given by αC, where $\alpha < 1$, then the total parasitic capacitance can be given by:

$$C_{par} = C_{fixed} + \alpha . C_{\max} \tag{4.2}$$

Now the ratio of f_{max} to f_{min} becomes:

$$\frac{f_{\max}}{f_{\min}} = \sqrt{\frac{C_{\max} + C_{par}}{C_{\min} + C_{par}}} \tag{4.3}$$

Hence, the presence of the parasitic tank capacitance reduces the ratio of f_{max} to f_{min}. Note that in the limit that $C_{par} \gg C_{max}$, this ratio tends to one. This means that for larger C_{par}, f_{max} becomes very close to f_{min}, severely limiting the tuning range. This becomes specially important at millimeter-wave frequencies, since the the tank capacitance is intrinsically low (due to the high frequency requirement), making

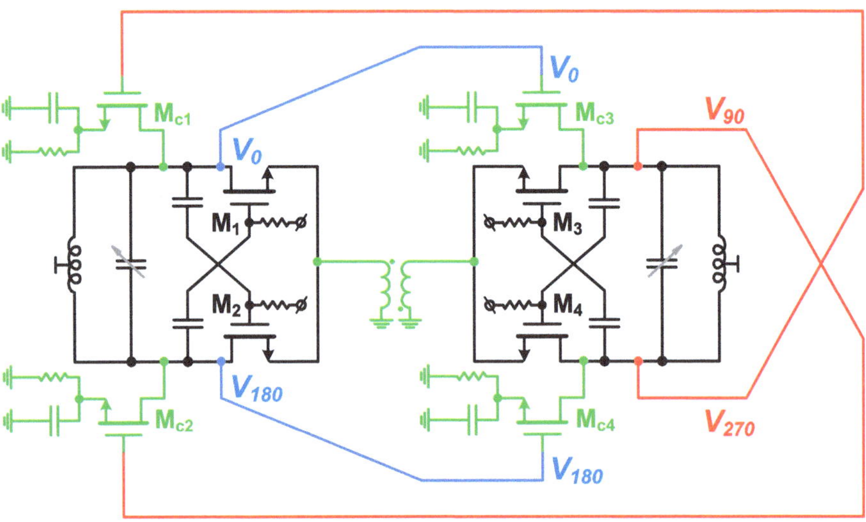

Fig. 4.1 Schemtic of the proposed millimeter-wave QVCO

the tuning range more sensitive to parasitics. It is, therefore, of utmost importance that the quadrature coupling technique used adds as little parasitic capacitance as possible.

Based on the previous discussion, the proposed QVCO architecture is shown in Fig. 4.1. To reduce parasitic loading, super-harmonic transformer coupling at the source-node is used as explained in Sect. 2.2.2. Since the transformer based super-harmonic coupling technique requires larger QVCO swings [25], it is more effective at the higher frequency bands where the amplitude is larger. To ensure proper coupling at the lower frequency bands, RC phase-shifting with small coupling transistors is used to assist the coupling operation at lower amplitudes (lower frequencies). The coupling transistors are almost seven times smaller than the cross-coupled g_m transistors, adding very small parasitic loading. The gates of the cross-coupled g_m transistors are AC-coupled, with separate gate biasing allowing a larger output-swing before the g_m transistors go into the linear region, and hence improving phase-noise performance [49].

Besides the parasitics from the coupling transistors, the capacitor bank parasitics become important, specially when a large tuning range is desired. Fig 4.2 shows one slice of the switched capacitor bank. Note that there are parasitic capacitances associated with both the top (C_{top})and bottom (C_{bot}) plates of the capacitor. There are also parasitics originating from the MOS switch (C_{sw}). When the switch is off, the top plate, bottom plate, and switch parasitics appear in parallel (assuming $C \ll C_{top,bot,sw}$). This lowers the maximum oscillation frequency as it increases the tank capacitance at the highest frequency setting (when all bank capacitors are disconnected). On the other hand, when the switch is on it has a finite resistance r_{on}.

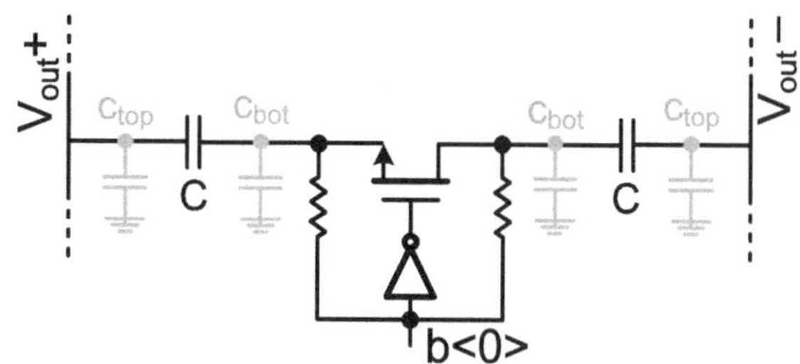

Fig. 4.2 One bit of the capacitor bank, and associated parasitics

This finite resistance puts an upper limit on the capacitor's quality factor, thus reducing the quality factor of the tank for lower frequencies, and *increasing* the lower frequency limit. *Hence, in the capacitor bank the capacitor and switch parasitics limit the upper frequency limit whereas the switch on-resistance restricts the lower frequency limit.*

4.3.1 Prototype Design

Based on the architecture proposed in Fig. 4.1, a prototype was designed in IBM's 32nm SOI process. Only discrete capacitor tuning is used, with 6-bits of binary-weighted tuning: a 3-bit coarse-tuning bank with a 15 fF unit capacitor, and 3-bit fine-tuning bank with a 3 fF unit capacitor. The high-Q capacitor available in the process is a Metal-Oxide-Metal (MOM) capacitor, which is formed by inter-digitated fingers of metal layers. The drawback of the MOM capacitor is its large parasitic capacitance, which is equal for both top and bottom plates. Parasitics are large due to the use of metal layers close to the substrate. They are also equal on both top and bottom plates due to the structure used as shown in Fig. 4.3a. This MOM capacitor uses three metal layers: Metal-6 through Metal-8.

To reduce parasitics, a custom capacitor is designed as shown in Fig. 4.3b. The custom capacitor employs only Metal-8 and Metal-9; these metals are further away from the substrate reducing the parasitic component. Furthermore, significant reduction in top plate parasitic is achieved by keeping the interdigitated structure only on the upper metal layer. The lower metal layer is a single metal sheet, with no interdigitation. The capacitor's top plate thus has lower parasitics to ground than the bottom plate, since the bottom plate shields the top plate from the substrate. The drawback, however, is a lower capacitance density resulting in larger capacitor area for the same capacitor value. The two unit capacitors (15 fF and 3 fF) are designed and simulated using 3D electromagnetic simulation through Integrand's EMX® EM-simulation tool [50].

The tail transformer is designed as a two-turn symmetric inductor with a ground connection at the center tap, as shown in Fig. 4.4. Due to the physical layout constraints, relatively long leads have to be used. The whole structure (transformer + leads) is EM-simulated, and is designed such that it resonates with the capacitance

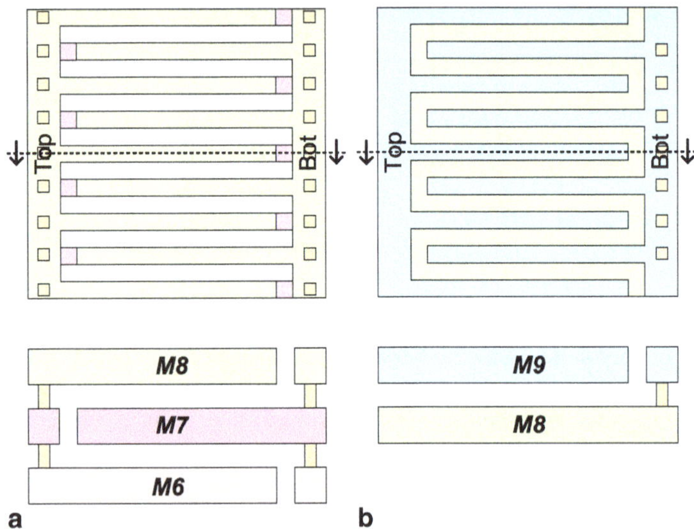

Fig. 4.3 Capacitor structures (**a**) MOM and (**b**) Custom

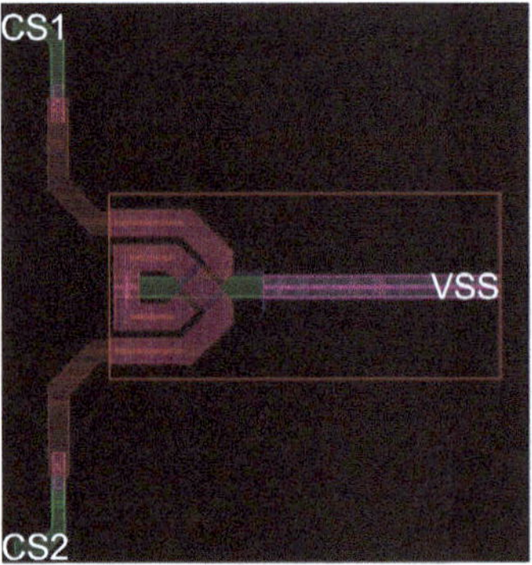

Fig. 4.4 Tail transformer layout

at the common sources of each of the two oscillators forming the QVCO (CS1 and CS2 nodes), with the resonance occurring at double the oscillation frequency [25].

The tank inductor is a symmetrical single-turn inductor with a differential inductance of 60 pH. Due to the large size of the capacitor bank (caused by both large capacitor areas, and a large number of bits), the interconnects add significant inductance. Hence, the tank inductance and the capacitor array are EM-simulated as a single structure to ensure the most accurate results. The full layout of the QVCO is shown in Fig. 4.5. The g_m-cell for each VCO is placed as close as possible to the tank inductor. Smaller capacitors are placed closer to the g_m-cell, and the larger capacitors are placed further away. This improves the tuning range by keeping the "effective" capacitance of the smaller capacitors in the capacitor bank almost unchanged (small lead inductance), while increasing the "effective" capacitance of the larger ones [51].

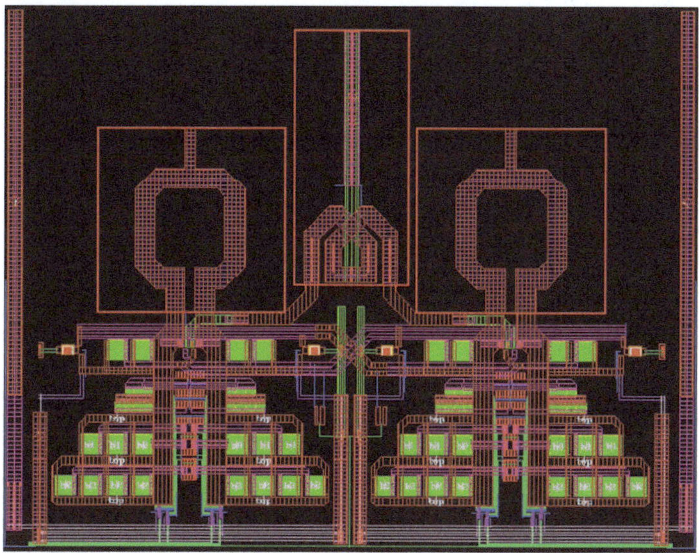

Fig. 4.5 Full layout of the proposed QVCO

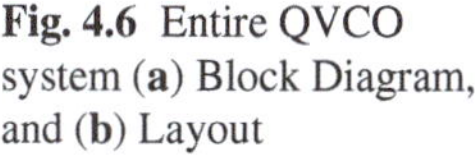

Fig. 4.6 Entire QVCO system (**a**) Block Diagram, and (**b**) Layout

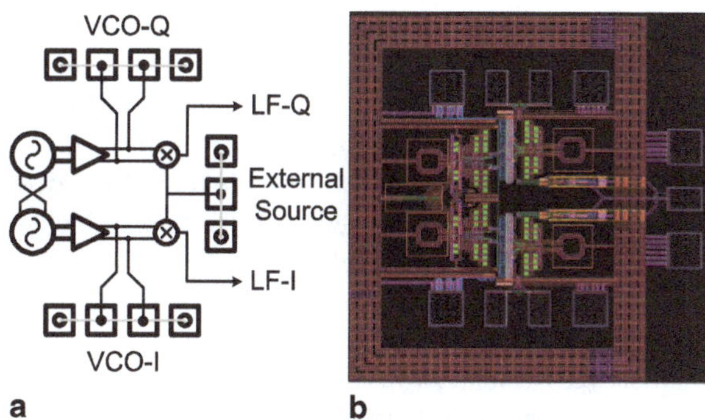

The QVCO is buffered using 50 Ω buffers, and the I and Q outputs are downconverted using mixers with an external input. The 50 Ω buffers are made up of the same tank as the QVCO, with only coarse tuning retained. A cascoded g_m-cell is used at the input, and a differential 100 Ω resistor is used for 50 Ω matching. The buffer outputs are connected to GSSG probes for on-chip probing. A GSG pad is used for providing the external input. The block diagram of the entire system (QVCO+buffer+mixers) is shown in Fig. 4.6.

The QVCO is simulated with full EM-extraction of the tank structure, as well as the tail transformer. A supply of 1V is used, with 600 mV of gate-bias. The simulated oscillation frequency of the QVCO versus the tuning word is shown in Fig. 4.7. With 3-bits of coarse tuning, eight different tuning "bands" can be distinctively identified. Within each band, eight different discrete frequencies can be observed corresponding

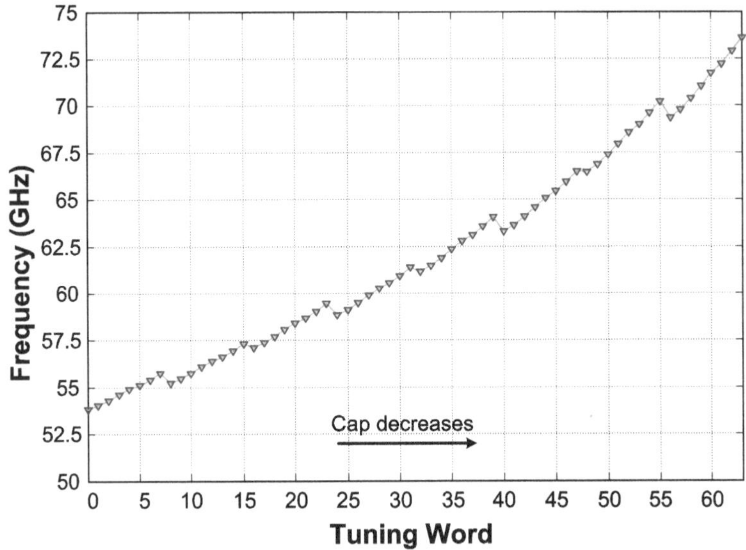

Fig. 4.7 Simulated output frequency versus tuning word

to the 3-bits of fine tuning. The QVCO can be tuned in 63 discrete tuning steps, ranging from 53.84 GHz to 73.59 GHz with a maximum frequency step size of around 700 MHz. *The QVCO has thus 19.75 GHz of frequency range, with a center frequency of 63.72 GHz corresponding to an FTR of 31 %.* This frequency range covers the 57–66 GHz band with ample room for process shifts. With a little shift in frequency (for instance via reducing the tank inductance), it can be made to cover the 71–76 GHz E-band as well.

The simulated power consumption of the proposed QVCO versus tuning word is shown in Fig. 4.8. The power consumption ranges from 23 mW at the highest frequency, to around 29 mW at the lowest frequency.

The phase-noise performance of the QVCO is also shown in Fig. 4.9, at 1 MHz and 10 MHz offsets from the carrier. The phase-noise at 1 MHz offset is comparable to the state-of-the-art performance in [24, 31, 46–48]. Except for [48], the QVCO's which have significantly better phase-noise have relatively low tuning ranges (maximum of 16.6 %) which allows for better phase-noise optimization. In [48], phase-noise at similar frequencies is very close to this work. Better phase-noise numbers are achieved at lower frequencies, as the design in [48] extends over the lower frequency range of 48.8–62.3 GHz. It is also to be noted that the phase-noise value at 10 MHz offset in Fig. 4.9 is almost 30dB lower than the value at 1 MHz offset, indicating a $\frac{1}{f^3}$ noise profile. This can be attributed to the relatively small transistor sizes used in the g_m-cell.

The figure of merit (FOM) of the QVCO, for phase-noise at 1 MHz and 10 MHz offsets, is shown in Fig. 4.10. At 1 MHz offset, the FOM ranges from 165 to 173 dBc/Hz. While the 173 dBc/Hz FOM benchmark is lower than (or equal to)

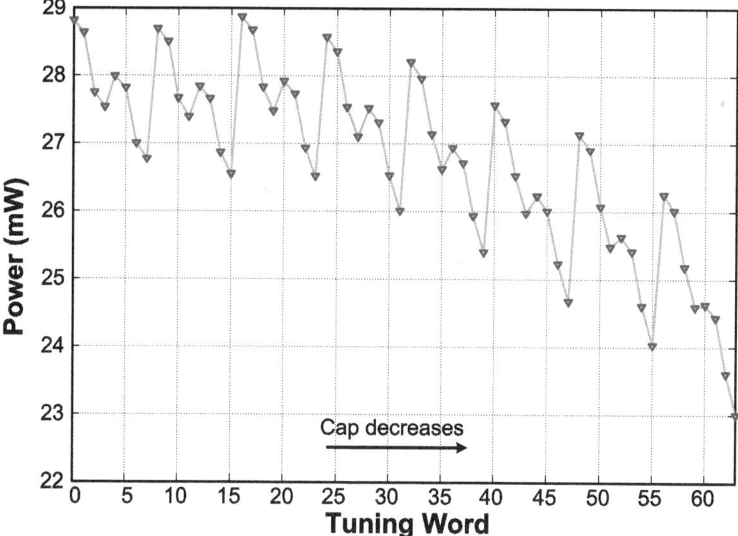

Fig. 4.8 Power consumption of the proposed QVCO versus tuning word

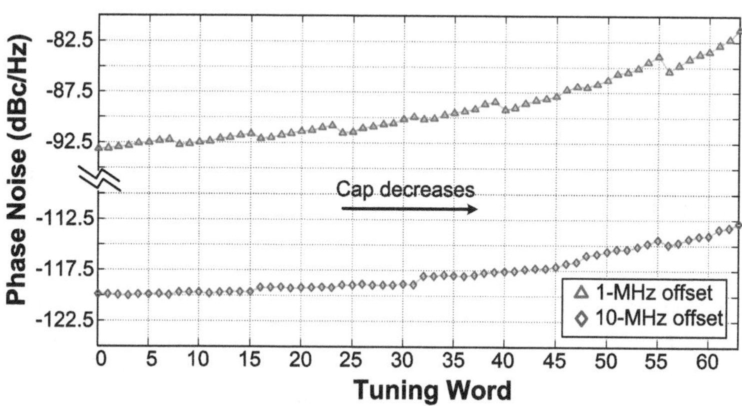

Fig. 4.9 Phase-noise of the proposed QVCO

the FOM achieved in [24, 31, 46–48], the achieved 19.75 GHz tuning range (and the corresponding 31 % FTR) is larger resulting in an FOMT of 183 dBc/Hz which is at par with state-of-the-art. The 10 MHz offset FOM ranges from 176 dBc/Hz to 180 dBc/Hz, with a corresponding FOMT of 190 dBc/Hz.

Table 4.1 shows a performance comparison of this work with state-of-the-art millimeter-wave QVCO designs. Amongst the designs presented in this comparison, the QVCO proposed in this work achieves the highest center frequency, and the highest tuning range (in absolute and relative terms) while achieving competitive phase-noise, FOM and FOMT numbers. Moreover, the presented design is the only

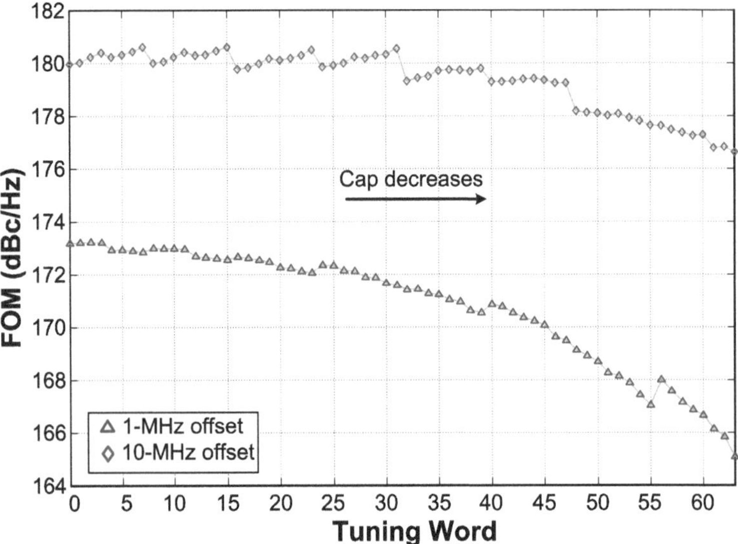

Fig. 4.10 Figure of merit of the proposed QVCO

Table 4.1 Performance comparison

Ref	f_o (GHz)	FTR %	PN(dBc/Hz)	FOM(dBc/Hz)	FOMT(dBc/Hz)	Power (mW)
[24]	63.1	16.6	−95	179	184	11.4
[31]	58.2	7.5	−97	179	177	22
[48]	55.5	24.3	−94	176	184	15.6–30
This Work	63.7	31	−92.5	173	183	23–29

one that covers the entire 57–66 GHz band, while also covering a considerable part of the 71–76 GHz E-band. In fact, with modification of the tank inductance the proposed QVCO can be made to simultaneously cover the whole 57–66 GHz, as well as the whole 71–76 GHz E-band, a feature not achieved by the other presented work.

4.3.2 *Measurements*

The micrograph of the fabricated chip is shown in Fig. 4.11. The three large inductors are *not* a part of the design. Three different prototypes of the QVCO were placed on chip, two with 6-bits of tuning and different placement of the g_m-cell, and a third one with only 3-bits of coarse tuning.

Although the three designs were simulated using full 3D EM-simulation, neither of the oscillators started up in actual measurements. The most likely reason is the

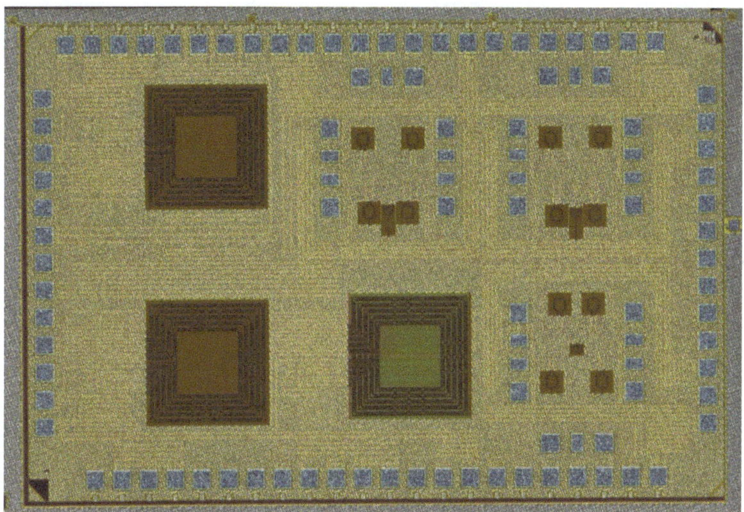

Fig. 4.11 Chip micrograph

lack of RF models in this specific process; the SOI process files available were mostly tailored for a digital/mixed-signal design flow. These lacked the essential modeling to account for the high-frequency non-quasistatic effects that would limit the performance at millimeter wave frequency. No capability was present to generate the high frequency models from measurements, and hence, the process files provided by the fab were used. Moreover, unexpected metal fill was performed by the fab on the top-most metal layer in the chip; this is a metal layer that is higher than the actual top-most metal used in the design. The effects of this metal fill had not been taken into account in the actual design (they mostly lead to increased loss and reduced quality factor).

Chapter 5
Quadrature Injection Locking for Channelized Receivers

5.1 Introduction

Wireless technology is continually demanding higher data rates. Increasing the bandwidth is the most effective way to achieve that goal [52]. Higher bandwidths, however, imply a high ADC clocking speed as well as a high dynamic range making the ADC power-hungry and difficult to design, if at all feasible. Frequency channelization is an effective way to overcome this problem. It reduces the required clock speed of the ADC and also increases the immunity to narrow-band interferers, hence reducing the ADC dynamic range requirements [4]. Channelization also reduces the number of independent in-band signals, hence, reducing the peak to average power ratio (PAPR) [53], and consequently an overall reduction in power consumption.

A channelized receiver, capable of achieving 4-GHz of instantaneous bandwidth, around a 21 GHz center frequency, is shown in Fig. 5.1. An external wideband low noise amplifier amplifies the entire band. The wideband signal is then down-converted into two 2-GHz channels using two sets of quadrature mixers, operating at 20 and 22 GHz. Two sets of lowpass filters, folloowing the mixers, complete the channelization. The desired wideband signal can then be reconstructed digitally by upsampling each digitized stream by two, followed by the use of digital reconstruction filters to equalize the filters' phase and amplitude responses [4]. For a faithful reconstruction of the original signal, the two channels need to be phase synchronous.

A major challenge for the receiver in Fig. 5.1 is the generation of two phase synchronous quadrature LOs at 20 and 22 GHz. PLL-based solutions such as [11] are ineffcient at mm-wave frequencies due to the need of high frequency dividers and SSB mixers, which are power-hungry . Besides, parasitics become prominent at these frequencies, limiting mixer linearity and resulting in a large number of spurious components.

Injection locking is an alternative approach for LO generation in mm-wave designs [42, 54–56]. Most designs, however, are aimed at a single LO frequency. The design in [54] implements simultaneous LOs but with two drawbacks: it doesn't generate quadrature signals and it only generates integer multiples of the reference signal. Hence, a different approach needs to be used for a more generalized solution [21].

M. Elbadry, R. Harjani, *Quadrature Frequency Generation for Wideband Wireless Applications,* Analog Circuits and Signal Processing, DOI 10.1007/978-3-319-13788-9_5

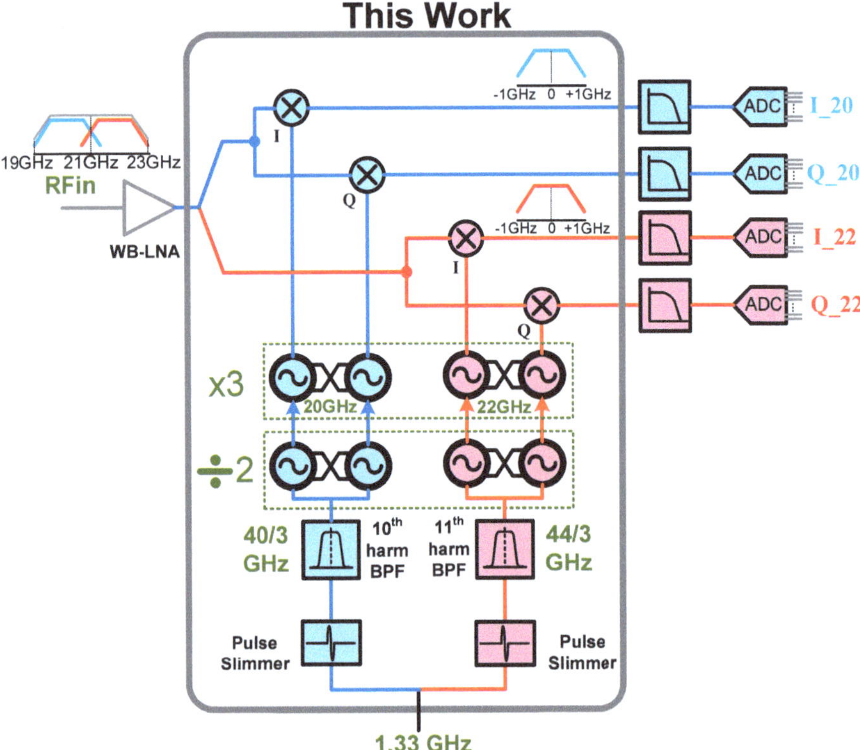

Fig. 5.1 Dual band 19–23 GHz channelized receiver

This chapter addresses the use of injection locking to simultaneously generate 20 and 22 GHz phase synchronous quadrature LOs using a single 1.33 GHz reference [57]. Section 5.2 provides an overall system overview. Section 5.3 presents circuit design details and simulation results. Section 5.4 discusses the EM design methodology used to ensure proper operation of high frequency oscillators. Section 5.5 provides measurement results. Finally, Sect. 5.6 outlines overall conclusions.

5.2 System Overview

The LO system block diagram is shown in Fig. 5.1. For this specific design, only two channels are implemented. The system, however, can be extended to an arbitrary number of channels. The major limiting factor would be the diminishing amplitude of higher harmonics of the pulse-slimmer for a given process technology, as well as larger chip area. Both even and odd harmonics of the input signal are used, unlike the designs in [16, 58, 59], increasing the flexibility and allowing smaller channel spacing.

The nth and $(n+1)^{th}$ harmonics of the reference are generated using pulse-slimmers. The pulse-slimmers' outputs contains a large number of undesired harmonics as well. The pulse-slimmer in each chain is, thus, followed by a band-pass filter (BPF) to emphasize the harmonics of interest and suppress the undesired ones. Up to this point, only a single phase is present. To generate quadrature outputs, the BPF's output is then fed to an injection locked frequency divider (ILFD) which performs a divide-by-2 operation. The quadrature outputs are frequency-multiplied using quadrature injection locked frequency multpliers (ILFM), generating $\frac{3}{2}nf_{ref}$ and $\frac{3}{2}(n+1)f_{ref}$ LOs. To generate 20 and 22 GHz, $n = 10$ and $f_{ref} = 1.33$ GHz are chosen. System aspects and architectural choices of each block are discussed further in the following subsections.

5.2.1 Pulse Slimmer

Each chain in Fig. 5.1 starts with a pulse-slimmer which generates all the harmonics of the input signal. The duty cycle of the pulse-slimmed signal is optimized in order to maximize the desired harmonics. For a square wave with a duty cycle D and a unity peak-to-peak amplitude, the amplitude y of the nth harmonic can be given by [60]:

$$y = D\frac{sin(n\pi D)}{n\pi D} \tag{5.1}$$

For the nth harmonic, there are several values of D which maximize the amplitude. According to Eq. (5.1) the absolute value of the amplitude of the nth harmonic is periodic in D with a period $D = \frac{1}{n}$. Hence, an additional criterion is needed to select the optimum duty cycle. In this case, it is desirable to reduce the amplitude of the lower harmonics (which is naturally larger) as well. This reduces the amount of desensitization to the following stage (the BPF), caused by the undesired low-frequency (and high amplitude) harmonics.

Based on these two criteria, a plot is made for the amplitude of the 10th harmonic and the ratio of the 10th harmonic to the fundamental versus D. As shown in Fig. 5.2, the maximum value of the 10th harmonic is periodic in D with a period of $\frac{1}{10}$. Nevertheless, the ratio of the 10th harmonic to the fundamental is higher for lower values of D. The absolute maximum of the ratio, which simultaneously corresponds to a maximum of the absolute value of the harmonic, occurs for $D \approx \frac{1}{20}$. This value of D is too small to be realized practically, hence the next higher value (highlighted in Fig. 5.2) is chosen. In a practical scenario, a value of D ranging from $\frac{1}{8}$ to $\frac{1}{6}$ can be used. This range of D is also usable for the other chain $n = 11$. This means that a single pulse slimmer can be used for both chains. However, in this design, chip floorplan considerations lead to the use of two pulse-slimmers (one for each chain) as discussed in Sect. 5.3.5.

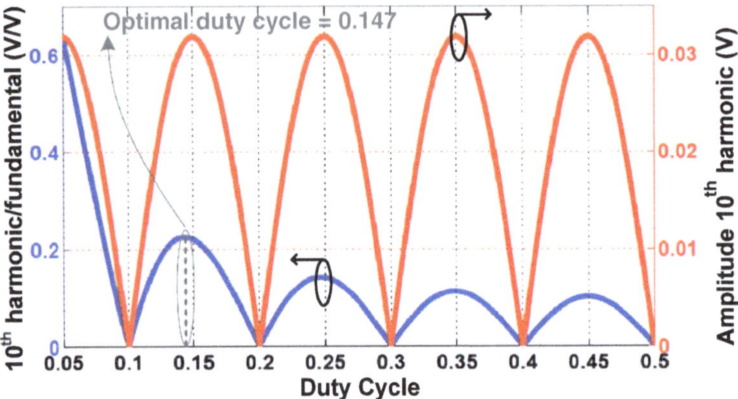

Fig. 5.2 Amplitude of 10th harmonic and ratio of its amplitude to the fundamental's

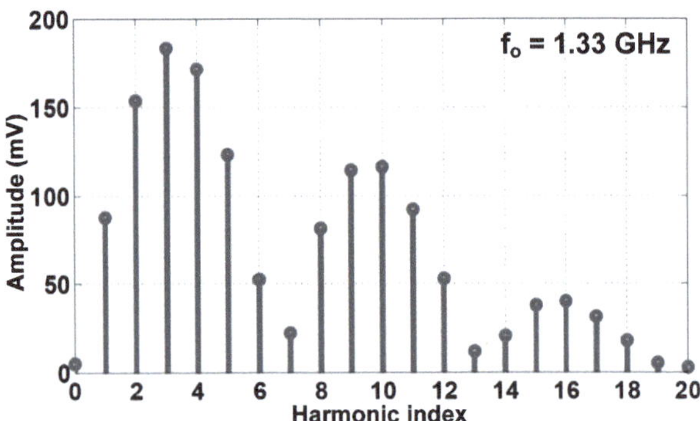

Fig. 5.3 Output spectrum of the pulse slimmer (SpectreRF® simulation)

In addition to duty cycle optimization, a differentiator is added after the pulse slimmer to further enhance higher harmonics and suppress lower ones. The circuit implementation details of the slimmer and the differentiator are discussed in Sect. 5.3.1.

5.2.2 Bandpass Filter

The techniques discussed in Sect. 5.2.1 help reduce the amplitude of the undesired harmonics. Nevertheless, the amplitude of lower harmonics remains higher than the desired harmonics. Figure 5.3 shows the output harmonics of the pulse slimmer in the 20 GHz chain as predicted by SpectreRF® simulations. It can be seen that the harmonics close to the fundamental are 2–3 times higher than the desired 10th harmonic. In the absence of a BPF, the high-amplitude low-frequency harmonics would saturate the input g_m stage of the ILFD, desensitizing it with respect to the desired harmonic. This, in turn, results in a limited lock range and hence higher phase

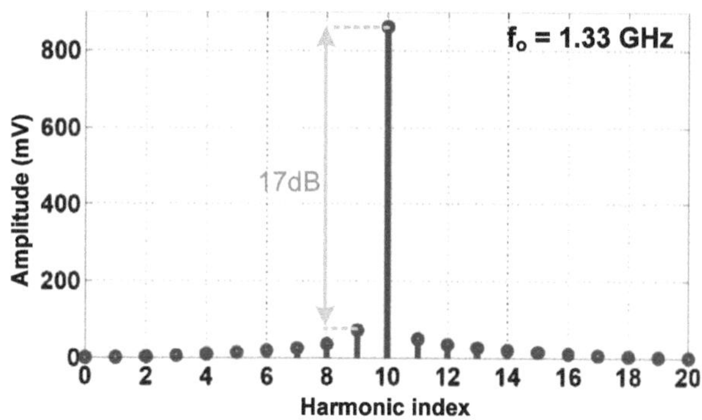

Fig. 5.4 Output spectrum of BPF(SpectreRF® simulation)

noise [61] and lower process tolerance. In the presence of the BPF, on the other hand, the low frequency harmonics are highly suppressed as shown in Fig. 5.4 (SpectreRF® simulations). An active injection-locking based BPF is used in this design which has the advantage of higher gain at the frequency of interest, as well as higher Q for the filtering action with a lower power consumption (the 10th harmonic is amplified by a factor of 7 and the highest harmonic is 17 dB below desired signal). The drawback, however, is the possibility for higher phase noise due to the intrinsic phase noise of the injection-locked oscillator (ILO). The output phase noise of a conventional BPF is approximately the same as the input's phase noise. With an injection-locked BPF, however, additional phase noise is added by the oscillating core. Nevertheless, with proper design (high enough lock range) the additional phase noise can be made negligible while retaining the high gain and high Q advantages [61]. A more detailed comparison between conventional and injection-locked based BPF is provided in Sect. 5.3.2.

5.2.3 *Injection Locked Frequency Divider*

Following the BPF, an injection-locked frequency divider similar to the one in [16] is used. The injection method, however, is different than the one in [16]. This is discussed in detail in Sect. 5.3.3.

An ILFD is used for quadrature generation to get good phase accuracy [16, 62, 63] without the use of multi-stage polyphase filters (which are power-hungry and result in signal attenuation [13]). It is to be noted that the ILFD in each chain operates at one third the final LO frequency. A similar divide-by-2 scheme for quadrature generation in a PLL would have required operation at *twice the final LO frequency*. The suggested scheme, hence, provides a more feasible solution with a lower power consumption.

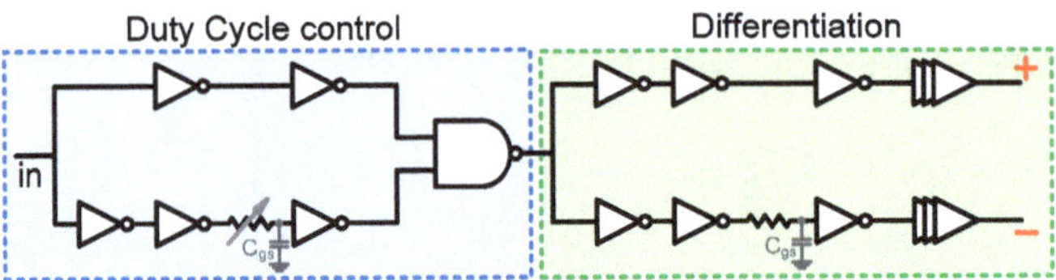

Fig. 5.5 Pulse slimmer (PSLIM) circuit diagram

5.2.4 *Injection Locked Frequency Tripler*

The final stage in each chain is an injection-locked frequency tripler. The tripler takes I and Q injection inputs and generates I and Q outputs at three times the input frequency [64]. Circuit design details of the multiplier are presented in Sect. 5.3.4.

5.3 Circuits

In this section, the circuit details of each of the synthesizer chain blocks are discussed and relevant simulation results are presented.

5.3.1 *Pulse Slimmer*

One of the key features of the proposed synthesizer architecture is that it makes use of both the even and odd harmonics of the reference clock signal. As discussed in Sect. 5.2.1, the 10th and 11th harmonics have to be maximized while suppressing the other unwanted harmonics. This is achieved by a two-stage, semi-digital design as shown in Fig. 5.5.

The first stage is a duty cycle control stage; it converts the input 50 % duty cycle clock to the optimal value discussed in Sect. 5.2.1 $\left(\frac{1}{6} \text{ to } \frac{1}{8}\right)$. This is done by NAND'ing two paths with a one inverter delay difference. Since the required delay is slightly higher, additional fine delay is added by means of a MOS resistor. The resistor together with its load cap forms an R-C time delay. This delay can be controlled off-chip by a variable voltage. In this process, one inverter delay amounts to around 26 ps; the variable delay can be changed from 35 ps to 83 ps as shown in Fig. 5.6.

The second stage performs two functions: it acts as a discrete-time differentiator, and it converts the single-ended signal to a pseudo-differential one. This is done by splitting the signal into two paths with equal number of inverters. One of the two paths, however, has an additional small delay implemented by a fixed MOS resistor (the two output signals are shown in Fig. 5.7a). Hence, the pseudo-differential output takes the form $\left(1 - z^{-1}\right) s(t)$, which is a differentiated version of the input signal $s(t)$

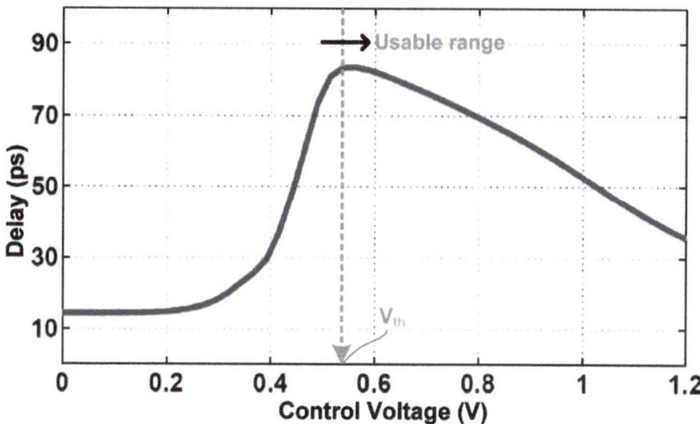

Fig. 5.6 Delay versus control voltage

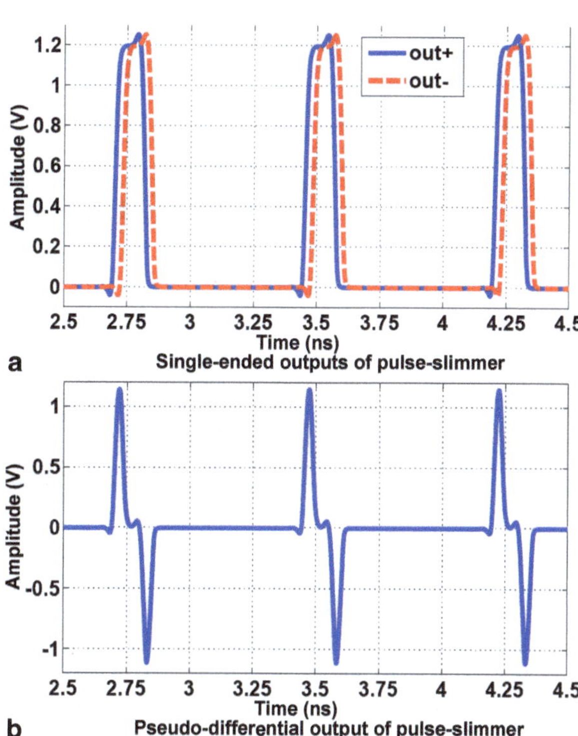

Fig. 5.7 Output of the pulse-slimmer. **a** Single-ended outputs of pulse-slimmer, **b** Pseudo-differential output of pulse-slimmer

(here the z^{-1} is the delay difference between the two paths). Combining this with the inherent low-pass nature of the chain results in an overall bandpass response as evident from Fig. 5.7b.

5.3.2 Bandpass Filter

As discussed in Sect. 5.2.2, an active injection-locking based bandpass filter is used in this design. Figure 5.8 shows the schematics of a conventional LC-based bandpass

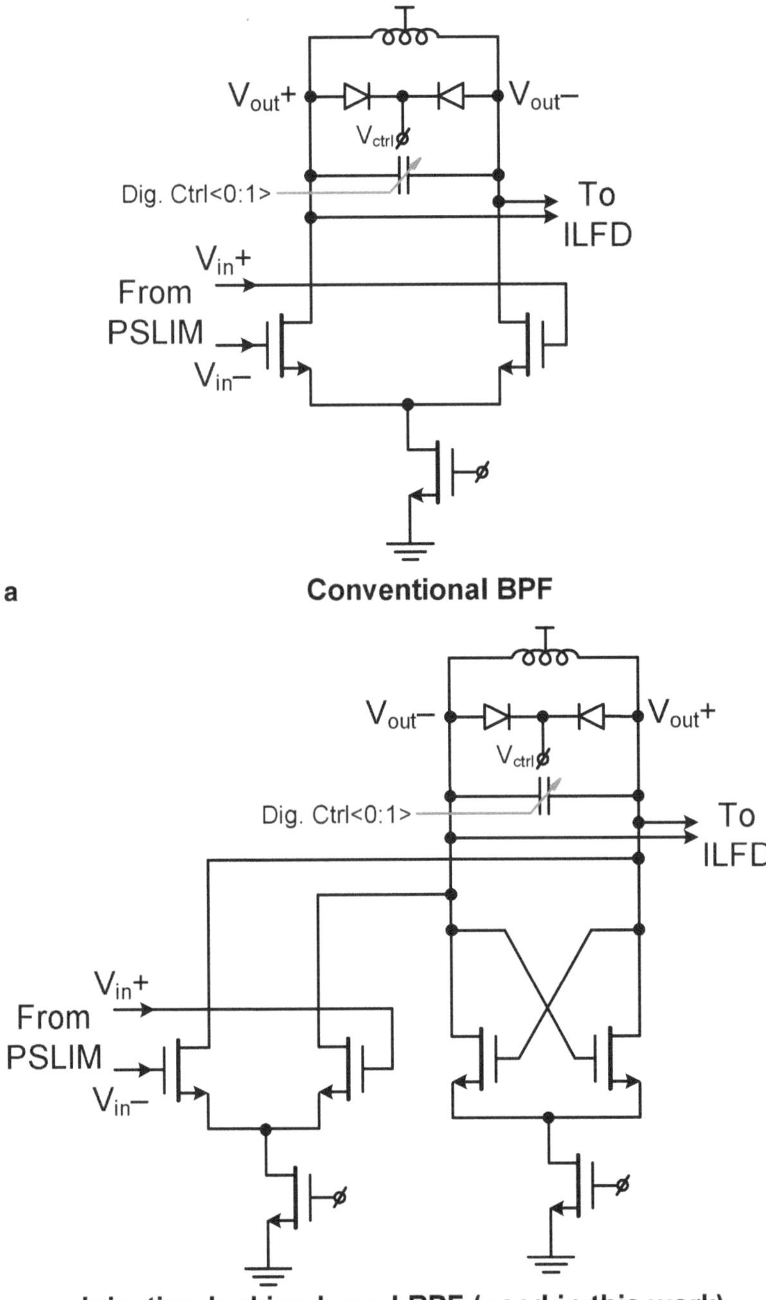

Fig. 5.8 Schematic of the bandpass Filter (*BPF*)—bias details omitted. **a** Conventional BPF, **b** Injection-locking based BPF (used in this work)

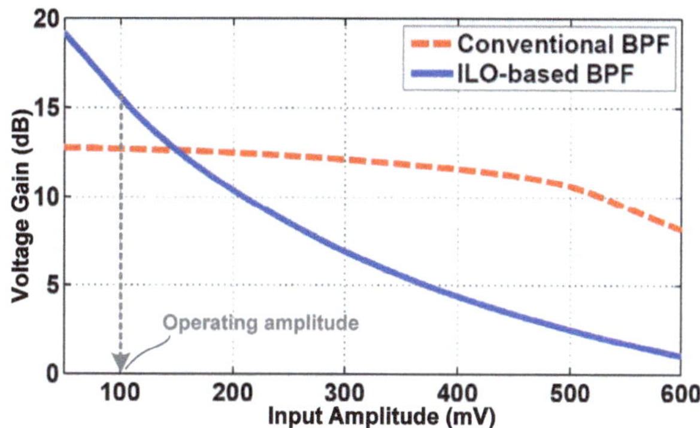

Fig. 5.9 Simulated gain of conventional and ILO based BPF (under equal power consumption)

filter and the injection-locking based bandpass filter used in this work. Under free-running conditions, the ILO's amplitude (V_{osc}) has a finite value. Once locked, the output amplitude remains the same creating an effective gain of V_{osc}/V_{inj}, V_{inj} being the injected signal's amplitude. With a constant output amplitude, the filter's gain depends on V_{inj} with smaller V_{inj} resulting in larger gain, as shown in Fig. 5.9. This nonlinear behavior, though undesirable for general filtering applications, doesn't pose a problem here as the desired signal is a single-tone sinusoid. As evident from Fig. 5.9, the conventional BPF has a flat gain versus input amplitude, whereas the ILO-based BPF's gain has a −6 dB/octave slope suggesting a $\frac{1}{V_{inj}}$ gain dependance. At a 100 mV input amplitude (the amplitude of the 10th harmonic from pulse-slimmer), the ILO-based BPF has a 3 dB higher gain than the conventional LC-based BPF. An additional benefit of the ILO based BPF is that the output amplitude saturates even for small inputs and remains relatively independent of the input signal amplitude, i.e., it behaves as an AGC providing constant amplitude to the next stage. The drawback, however, is that the phase noise increases at lower amplitudes [61], suggesting a moderate input level as a good compromise.

In addition to the higher gain, the injection-locking based bandpass filter also provides higher suppression of undesired harmonics. A spur at an offset f_m from the input signal is suppressed by f_m/f_L [58], where f_L is the single-sided lock range. As shown in Fig. 5.10, the effective Q of the ILO-based BPF is almost twice that of the passive LC tank ($Q \approx 30$ for ILO-based filter versus $Q \approx 13$ for conventional filter). Two ILO-based BPFs are designed for the 20 and 22 GHz chains, with center frequencies of 13.33 and 14.66 GHz respectively. Each ILO draws 4-mA from a 1.2 V supply (Fig. 5.11).

Varactor

Bias details of the varactor are shown in Fig. 5.12. The same structure is also used in the ILFD and the ILFM. Hyper-abrupt PN junctions are used as varactors due to their higher quality factor (than that of the conventional MOS varactors) at the frequencies of interest. As shown in Fig. 5.13, a large parasitic capacitor is formed between the varactor's negative terminal and the substrate degrading the Q-factor.

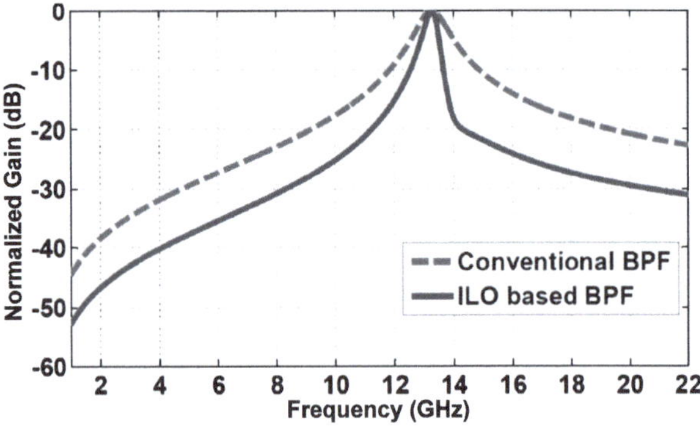

Fig. 5.10 Simulated quality factor of conventional and ILO-based BPF

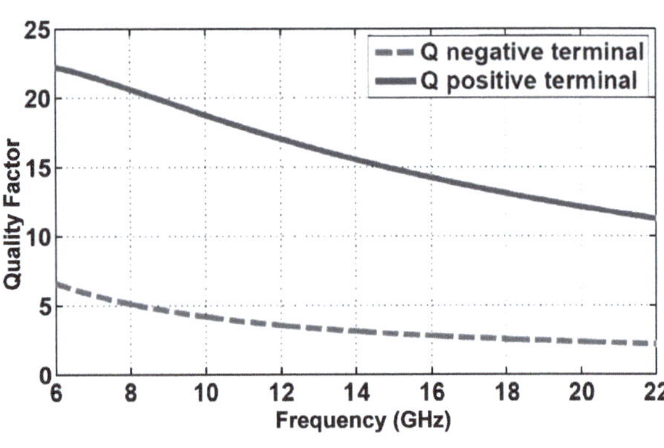

Fig. 5.11 Varactor quality factor measured at negative and positive terminals

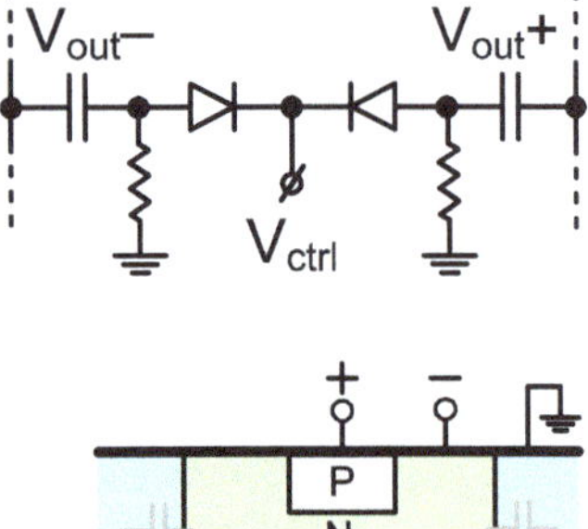

Fig. 5.12 Varactor bias details

Fig. 5.13 Structure of hyper-abrupt PN-junction varactor (parasitics highlighted)

The positive terminal, on the other hand, is not prone to this parasitic capacitance. Hence, the positive terminal has a Q-factor which is 3–5 times higher than that of the negative terminal as shown in Fig. 5.11.

Capacitor Bank

The details of the two-bit capacitor bank are shown in Fig. 5.14 [65]. A similar capacitor bank structure is used in the ILFD and the ILFM. When the switch control

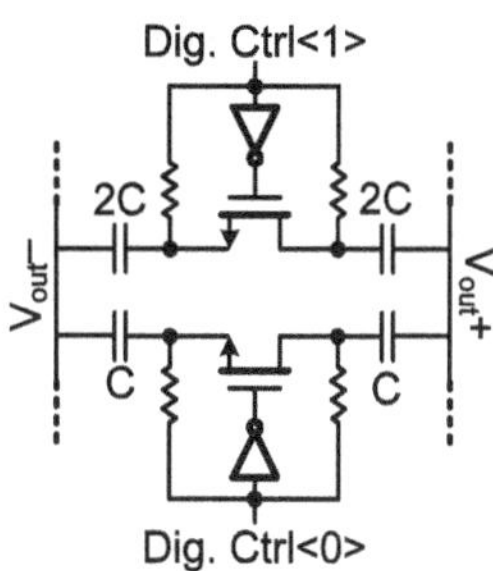

Fig. 5.14 Two-bit capacitor bank details

is at V_{dd}, the resistor bias is at ground and vice-versa. Without this bias scheme, an off-switch might turn on if the output swing is high. This can happen when the output voltage is close to its trough. Parasitic switch capacitance would cause the switch source voltage to spike down, in effect producing a net positive V_{gs} and causing the switch to turn on for a fraction of the cycle. By forcing V_{dd} bias through a resistor, this undesired effect is avoided.

5.3.3 Injection Locked Frequency Divider

As discussed in Sect. 5.2.3, an injection locked frequency divider is used in this work to generate quadrature outputs from the single phase reference. The design is based on the work in [16]. This, in turn, is similar to the architecture proposed in [66] but with inter-oscillator injection added to form a quadrature oscillator as the core (the core quadrature oscillator is based on [12]). Unlike [16] and [66], however, injection is not performed at the tail current source as shown in Fig. 5.15. Since no buffer is used between the BPF and the ILFD, injection at the tail source presents a large load capacitance to the BPF increasing its power consumption considerably. To avoid this, direct injection is used as suggested in [67]. The injected signal is applied to an NMOS switch in parallel with the oscillator's LC tank. In this work, the input has a 50 % duty cycle and hence the switch is on only once during a whole cycle of the injection signal. This, in effect, makes the output frequency half that of the input. The switch's input capacitance is roughly 25 times smaller than that of the tail current transistor, simplifying the design of the preceding BPF. Two ILFDs are designed, a 6.66 GHz ILFD for the 20 GHz chain and a 7.33 GHz ILFD for the 22 GHz chain; each consumes 27 mA from a 1.2 V supply.

It is to be noted that the DC bias at the NMOS switch terminals in Fig. 5.15 is V_{dd}. Hence it is not possible to turn the switch on unless the bias point is adjusted, which necessitates the use of AC coupling. Towards this end, both the gate connection and the source and drain connections of the switch are AC coupled. This provides more flexibility in choosing the bias point. The bias details of the switch are shown in Fig. 5.16.

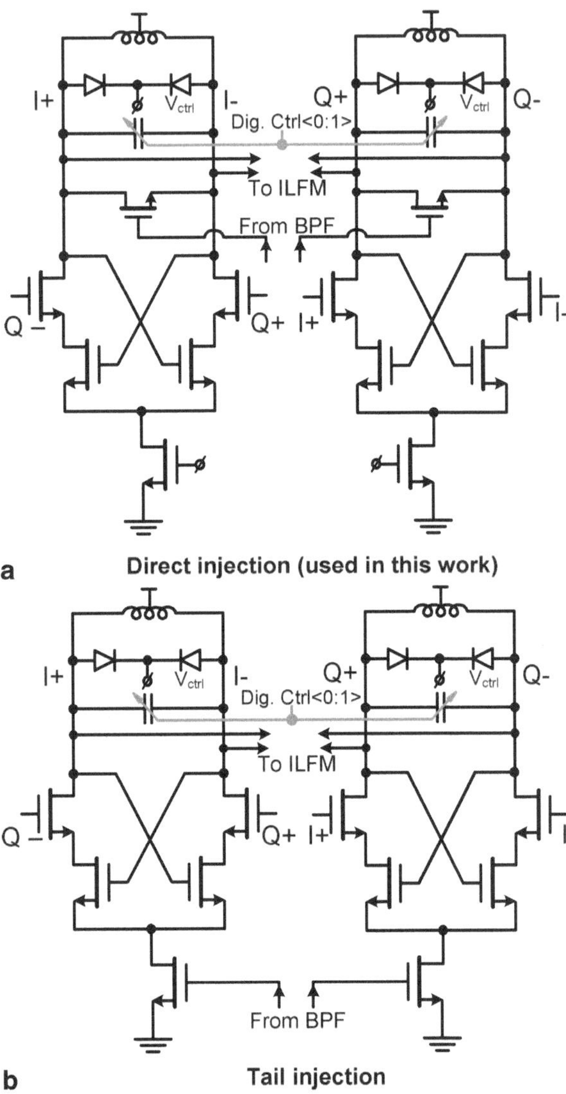

Fig. 5.15 Schematic of Injection Locked Frequency Divider (ILFD)—bias details omitted. **a** Direct injection (used in this work), **b** Tail injection

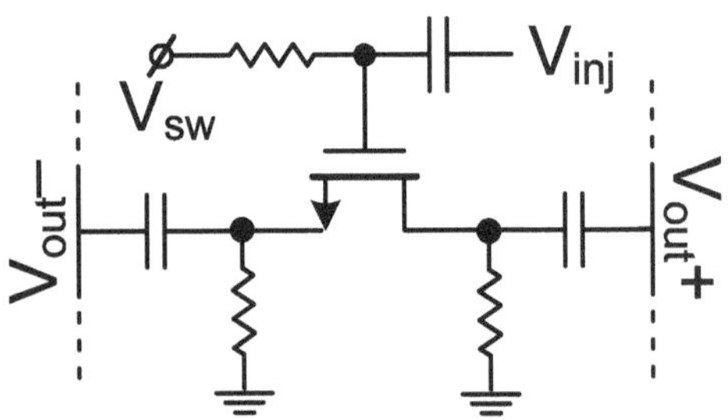

Fig. 5.16 Injection switch bias details

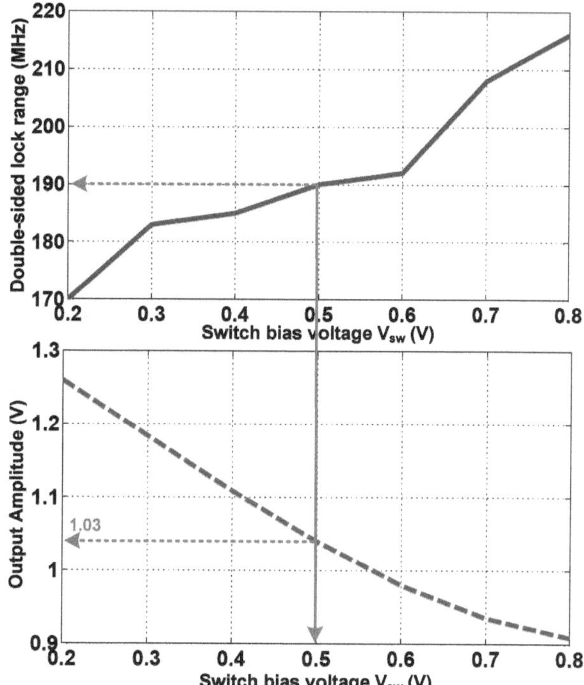

Fig. 5.17 Lock range and output amplitude versus switch bias for the ILFD

With the NMOS switch on, the quality factor of the tank is reduced. The time-averaged quality factor of the tank thus depends on the switch on-resistance, as well as the switch on time. The time-averaged quality factor, in turn, determines two important aspects of the ILFD: lock range and output amplitude. A wider lock range is desirable to suppress the ILFD's intrinsic phase noise and to cope with process variations. A higher output amplitude is also desirable to increase the lock range of the following stage, the ILFM. There is a clear trade-off between the two requirements, which is controlled by the switch's bias point. A higher bias voltage leads to a lower time-averaged quality factor and hence, a wider lock range and a lower output amplitude, and vice-versa. Figure 5.17 shows the lock range and the output amplitude of the ILFD versus the switch bias point V_{sw}. A reasonable trade-off is achieved for $V_{sw} \approx 0.5$ V, resulting in a 190 MHz double-sided lock range and a 1.03 V output amplitude.

Monte-Carlo simulations are performed to determine the expected phase-error and amplitude mismatch in the ILFD's output. As shown in Fig. 5.18, the simulated 3σ phase error amounts to $\pm 1.27°$ whereas the 3σ amplitude mismatch amounts to 0.12 dB.

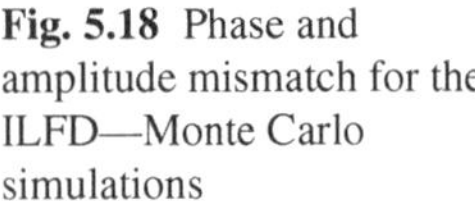

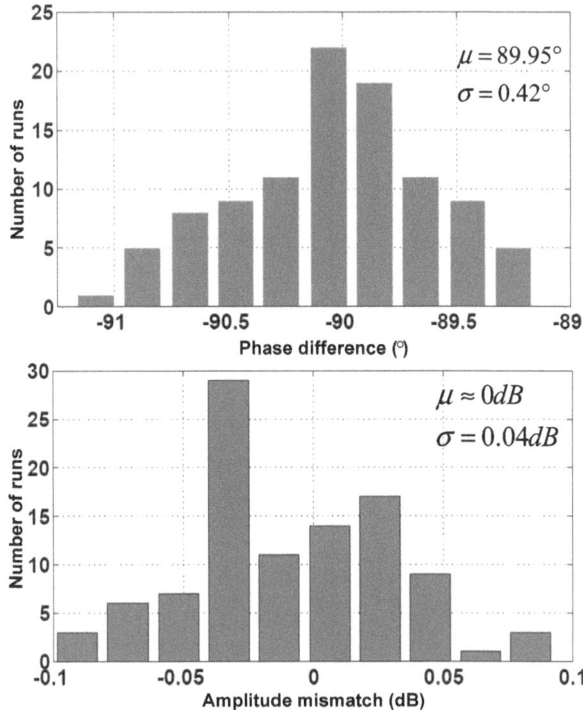

Fig. 5.18 Phase and amplitude mismatch for the ILFD—Monte Carlo simulations

5.3.4 Injection Locked Frequency Tripler

The injection locked frequency multiplier (ILFM), acting as a tripler, is shown in Fig. 5.19 [64]. At the core of the ILFM is a bottom-series-coupled quadrature QVCO similar to [68]. The injection differential pair is biased in subthreshold, and is operated in class-B mode resulting in a strong third harmonic current component. As shown in Fig. 5.20, the third harmonic of the injection current peaks below the threshold voltage V_{th}. An added advantage of subthreshold operation is the reduced power consumption in the injection pair. Two ILFMs are designed, with 20 and 22 GHz center frequencies. Each consumes 21 mA from a 1.2 V supply.

Monte Carlo simulations are performed to estimate the phase and amplitude mismatch of the ILFM. The results, shown in Fig. 5.21, represent *the intrinsic* phase and amplitude mismatches of the ILFM. The intrinsic 3σ phase error amounts to $\pm 2.28°$ whereas the 3σ amplitude mismatch amounts to 0.18 dB.

Since the final LO amplitude is large enough for full mixer switching, amplitude mismatch at the LO output is irrelevant. Phase mismatch, however, is citical. The total phase mismatch at the LO output is affected by the intrinsic phase mismatch of the ILFM, as well as the phase mismatch of the ILFD (which drives the ILFM). For a given fixed phase error of the ILFD $\Delta\phi_{in}$ and a fixed phase error of the ILFM $\Delta\phi_{int}$, a pessimistic estimate of the output LO phase error would be:

$$\Delta\phi_{LO} = \Delta\phi_{int} + 3 \times \Delta\phi_{in} \quad (5.2)$$

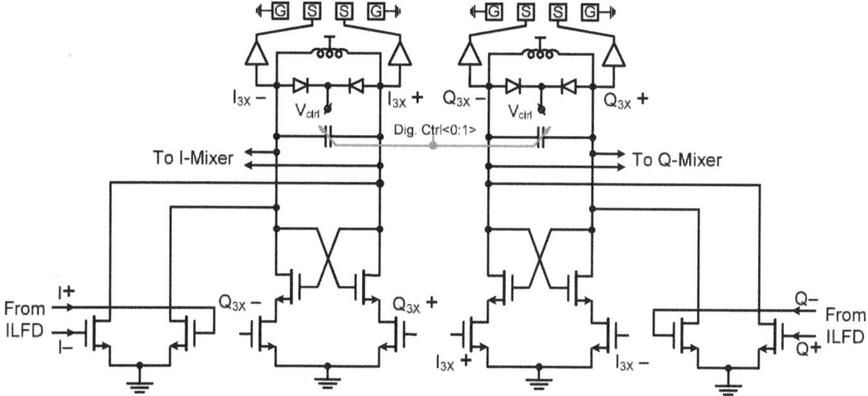

Fig. 5.19 Schematic of Injection Locked Frequency Multiplier (ILFM)—bias details omitted

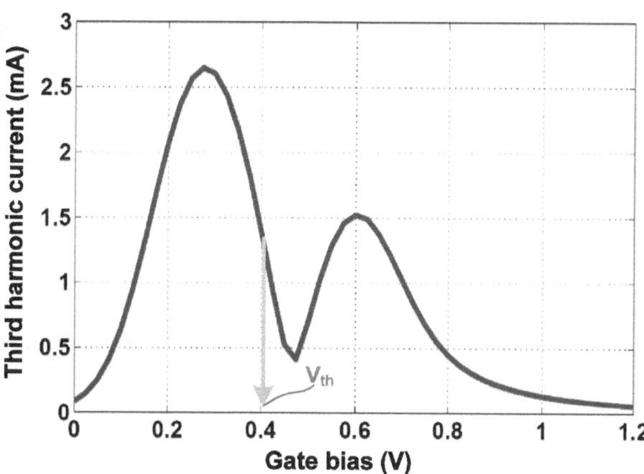

Fig. 5.20 SpectreRF® simulation—Third harmonic current in injection pair

Hence, the standard deviation of the phase error at the output will be the r.m.s sum of the standard deviations of the two terms in Eq. (5.2):

$$\sigma_{LO} = \sqrt{\sigma_{int}^2 + (3 \times \sigma_{in})^2} \tag{5.3}$$

Accordingly, the estimated 3σ phase error at the output of the LO is $\pm 4.4°$. This is a pessimistic estimate that represents an upper bound on the output phase error.

5.3.5 *Chip Floorplan*

Chip floorplanning is an integral part of the design process that is critical for successful operation. The floorplan of the chip in this work is presented in Fig. 5.22. The input reference clock signal is terminated on-chip via a 50 Ω resistor for matching.

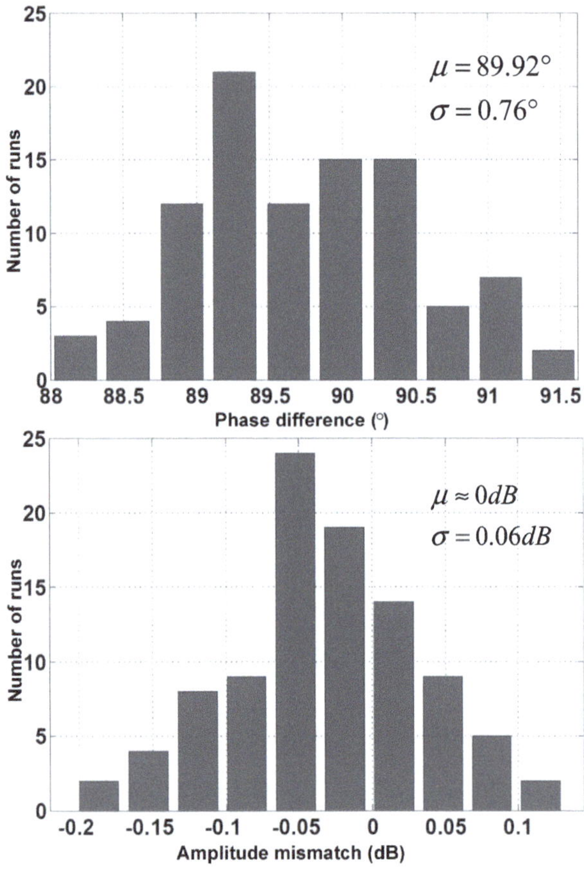

Fig. 5.21 Phase and amplitude mismatch of the ILFM- Monte Carlo simulations

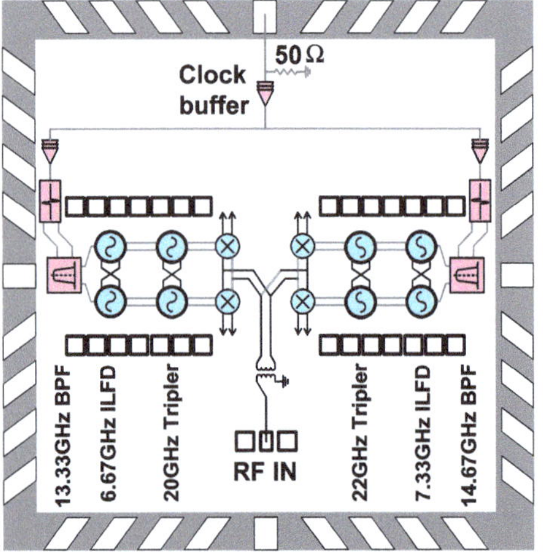

Fig. 5.22 Chip floorplan

The clock is buffered and then two separate paths are routed to the pulse-slimmer of each chain. Since the reference clock has a relatively low frequency, it is easier to route it for a longer distance. Another set of inverter buffers is also introduced along each path for further signal enhancement. As noted in Sect. 5.2.1, a single pulse-slimmer could be used. However, this would require its output to be routed for a very long distance. This would, in turn, impose two problems: attenuation of the desired high frequency harmonic content, and undesired parasitic coupling to different points in the chain. Hence, one pulse-slimmers is used for each chain. The slimmer is placed as close as possible to the bandpass filter to minimize the routing distance and hence, minimize signal loss and parasitic coupling. The associated area and power overhead are negligible. The output of the slimmer is not available for measurement and neither is the output of the bandpass filter. The injection locked divider and the injection locked multiplier follow in cascade. Care is taken to achieve maximum layout symmetry. The outputs, of both the divider and the multiplier, are buffered and fed to on-chip GSSG pads for probing. The final LO outputs are fed to active Gilbert-cell based double-balanced quadrature mixers for measurement purposes. The other input of the mixer is provided externally through probing via an on-chip GSG pad followed by an on-chip balun.

5.4 EM Design Methodology

In high frequency oscillators, interconnects play a critical role in the performance. Interconnects can no longer be treated as wires with a constant resistance (i.e. against frequency). Higher order effects kick-in, considerably altering the performance. The three major effects are: skin effect, substrate loss and current-crowding (or proximity effect) [69, 70]. If not accounted for, the extra AC resistance added by these effects can cause a significant degradation in the oscillation amplitude or, in the extreme case, a total startup failure. Moreover, with low tank inductances (150 pH $\sim$ 750 pH) the interconnect inductance cannot be neglected and can cause considerable frequency shifts. Due to the complex nature of these effects, as well as the complex interconnect pattern typical of integrated circuits, EM simulations become crucial for successful design validation. EM simulation tools, however, are not suited for extraction of the transistor parasitics. Moreover, as the number of simulated metal layers increases EM simulation times increase significantly.

To both accommodate transistor parasitics and speed-up EM simulations, a divide-and-conquer approach is used for extraction. Each oscillator layout is dissected vertically into two sections: a lower section (i.e. closer to substrate) which includes the transistors and the interconnects on the lower five metals and an upper section (i.e. farther away from substrate) including the inductors, capacitors and upper three metal interconnects. This choice is made because the high-frequency signals are routed on the top three metal layers (which are the thickest) to reduce DC resistance. Moreover since the thickness of the top three metals is larger than the skin depth, their resistance varies significantly at high-frequencies, as opposed to the thin

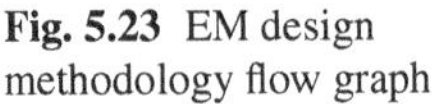

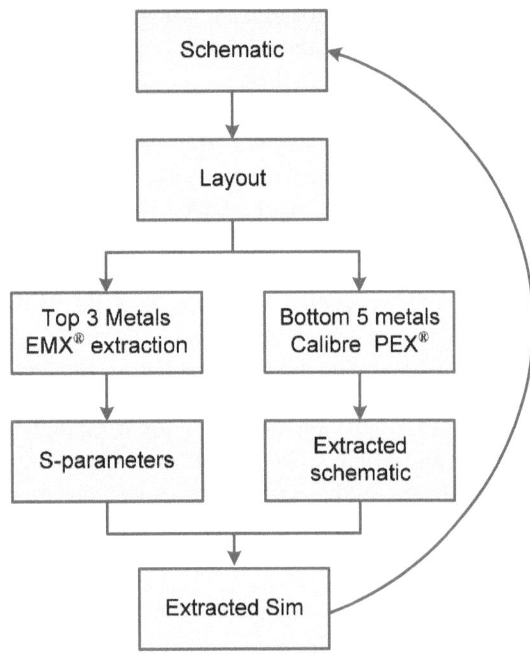

Fig. 5.23 EM design methodology flow graph

lower metals whose resistance is dominated by the DC value. Conventional circuit extraction tools (in our case Calibre PEX®) are used on the lower section resulting in an extracted schematic with lumped resistor and capacitor parasitics. For the upper section, EM simulations are used to capture parasitic AC resistances and parasitic inductances and capacitances. Due to the large interconnect structures typical of quadrature VCOs, a fast and efficient EM simulation tool was needed. Towards this end, Integrand's EMX® EM-simulation tool [50] was used. The tool outputs an S-parameter data file, which is then combined with the extracted schematic from the lower section to perform a simulation of the whole structure. The circuit is then modified, if needed, and another simulation iteration is performed until the desired performance (frequency/output amplitude) is obtained. Figure 5.23 summarizes the design methodology.

5.5 Measurements and Discussion

A prototype chip was fabricated in IBM's 130 nm CMOS technology. The chip micrograph is shown in Fig. 5.24. The active area is 1.8 mm^2. The test setup is shown in Fig. 5.25. A reference 1.33 GHz signal is supplied using an Agilent E8257D signal generator. An on-chip 50 Ω resistor provides termination for the generator. The quadrature oscillator outputs (for ILFD and ILFM)are measured using GSSG probes and the output is displayed on an HP E4407B spectrum analyzer where the

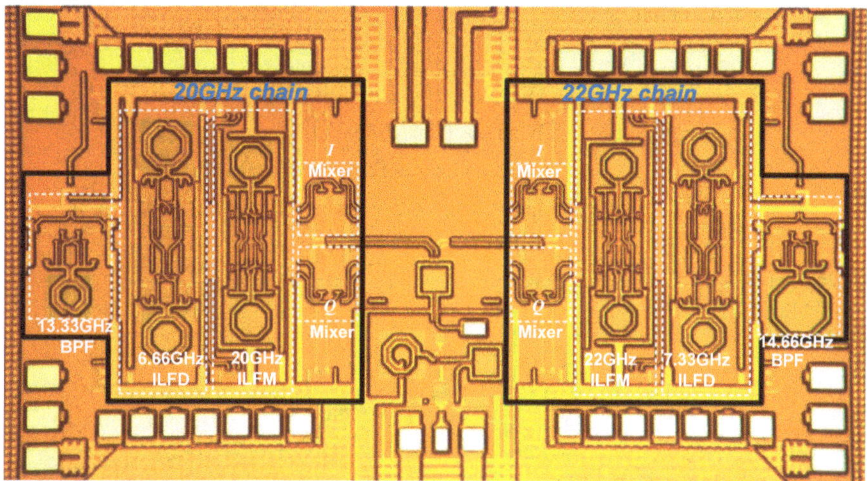

Fig. 5.24 Chip micrograph

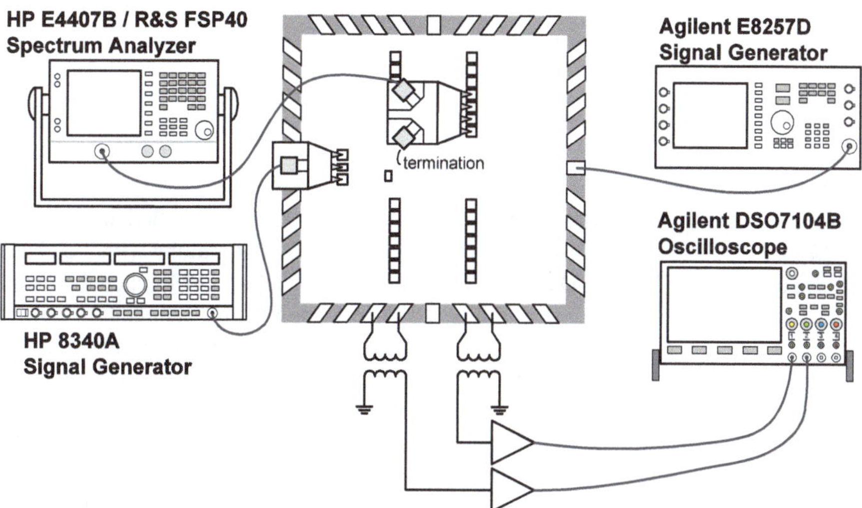

Fig. 5.25 Test setup

spur levels are measured. A R&S FSP40 spectrum analyzer is used to measure the phase noise. The GSSG probes are driven by on-chip 50 Ω buffers. The RF signal is supplied from an HP 8340A signal generator through GSG probes. The differential outputs of the quadrature downconversion mixers are connected to off-chip baluns. The single-ended outputs are then amplified and displayed on an Agilent DSO7104B oscilloscope.

Table 5.1 Current consumed per block

Block	Current (mA)
Digital (buffering + pulse slimmer)	3
Bandpass Filter (BPF)	4
Injection Locked Frequency Divider (ILFD)	27
Injection Locked Frequency Multiplier (ILFM)	21

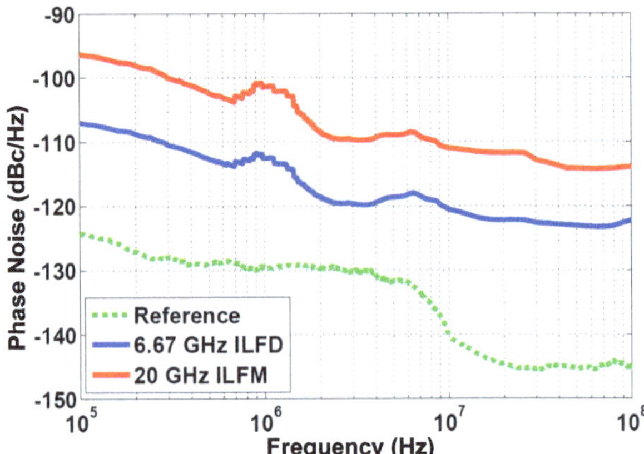

Fig. 5.26 Phase noise measured along the 20 GHz chain

Each chain consumes around 55 mA from a 1.2 V supply, for a total current consumption of 110 mA. The current consumed by each block is detailed in Table 5.1.

Figure 5.26 shows the phase noise of the reference oscillator along with the phase noises of the 6.67 GHz ILFD and the corresponding 20 GHz ILFM, when the chain is locked to the 1.33 GHz reference. The ILFD phase noise is measured at the output of the ILFD's 50 Ω buffers using GSSG probes. Similarly, the ILFM phase noise is measured at the output of the ILFM's 50 Ω buffers using GSSG probes as well. It is clear from Fig. 5.26 that the 20 GHz ILFM has negligible contribution to the output phase noise; the phase noise at the ILFM output is a faithful replica of the ILFD shifted by around +10 dB (since the ILFM multiplies its input frequency by three, it adds $20log(3)$ ≈9.5 dB to the input phase noise). This can be attributed to the relatively large lock range of the ILFM (single-sided lock range> 350 MHz). In contrast, the loop bandwidth of a PLL is typically in the range of a few MHz leading to a large contribution from the PLL's VCO to the output phase noise. The 22 GHz chain performs in a similar manner to the 20 GHz chain as shown in Fig. 5.27. The spot phase noise values at 1-MHz and 10 MHz offsets along the two chains are summarized in Table 5.2.

It is instructive to find the contribution of the reference, BPF and ILFD to the output phase noise. Towards this end, the model in [61] is used. Each ILO acts as a first-order low-pass filter to its input phase noise, with the corner frequency determined by the single-sided lock range f_L. The intrinsic ILO phase noise is

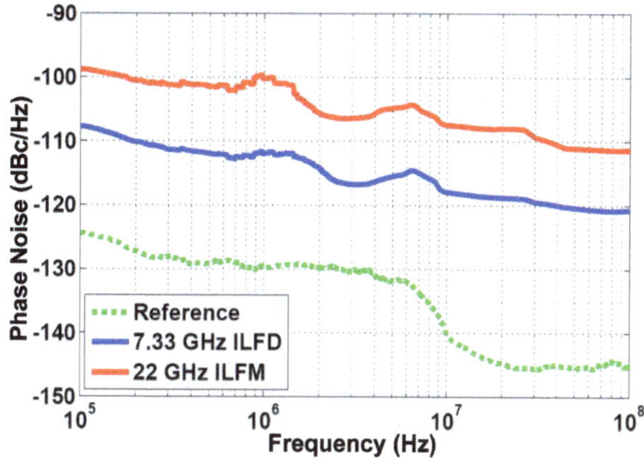

Fig. 5.27 Phase noise measured along the 22 GHz chain

Table 5.2 Phase noise performance

Offset frequency (MHz)	1	10
Measured @6.67 GHz (dBc/Hz)	− 112.5	− 121
Measured @20 GHz (dBc/Hz)	− 101.5	− 111
Measured @7.33 GHz (dBc/Hz)	− 112	− 118
Measured @22 GHz (dBc/Hz)	− 100	− 107.5

filtered by a first order high-pass filter of corner frequency f_L as well. Frequency division and multiplication scale the phase noise accordingly. The resultant model for the cascade of pulse-slimmer, BPF and ILFD is shown in Fig. 5.28.

Using this model, the output-referred phase noise contributions at the 6.67 GHz ILFD are plotted in Fig. 5.29 (BPF and ILFD phase-noise simulated, reference phase-noise measured). As evident from the figure, the BPF phase noise contribution is negligible. Even with the slight increase in actual BPF phase noise (due to modeling inaccuracies and supply-noise), BPF phase noise contribution remains negligible[1]. The simulated ILFD phase noise contribution is small but increases at larger offsets. The discrepancy between the simulated phase noise at the output of the ILFD (shown in Fig. 5.29) and the actual measured phase noise at the ILFD output (shown in Fig. 5.26) can be attributed to two reasons: (1) the actual ILFD phase noise is higher than simulated (again due to modeling inaccuracies and supply-noise) and (2) the measurement instrument intrinsic phase noise is considerable compared to the up-converted reference phase noise [71]. This results in a final phase noise profile that is different from the simulated curve.

The spurious performance of the 20 GHz LO chain is shown Fig. 5.30. The spurs are at 1.33 GHz offset; the significant harmonics at the BPF's output are the 9th, 10th and 11th harmonics. At the 6.67 GHz ILFD, the 10th harmonic is divided down

[1] The actual phase noise of BPF could not be measured because its output is not available for probing.

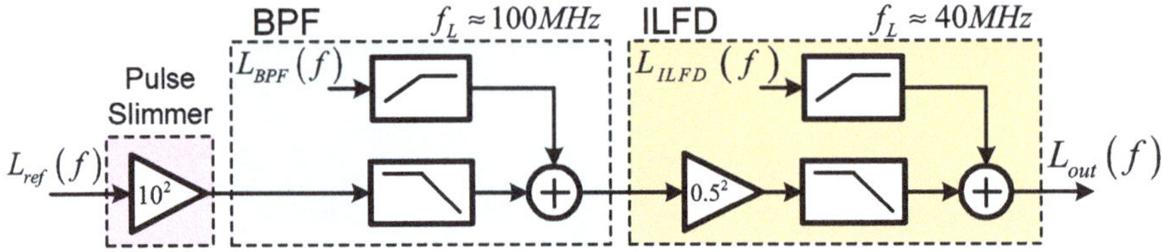

Fig. 5.28 Phase noise model

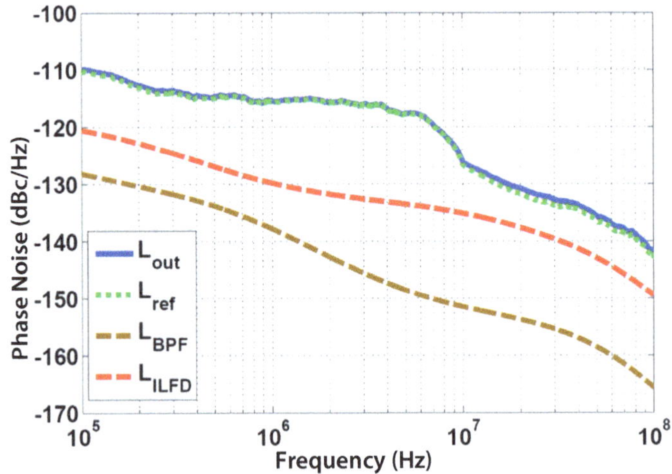

Fig. 5.29 Output-referred phase noise contributions at 6.67 GHz ILFD

generating the 6.67 GHz signal. The 9th and 11th harmonics, on the other hand, mix with the 6.67 GHz LO generating spurs at ±1.33 GHz offset. These spurs are suppressed by virtue of the ILFD's filtering effect (in a manner similar to the effect of the BPF). At the 20 GHz ILFM, the spur levels are further attenuated. The maximum spur at the output of the 20 GHz ILFM falls at −43 dBc as compared to −30 dBc at the output of the 6.67 GHz ILFD. While the BPF measurements are not available, the simulations shown in Fig. 5.4 suggest that the spurs at the BPF's output fall at around −17 dBc. Hence the cascade of ILOs provide excellent spur suppression.

Figure 5.31 shows the spurious performance of the 22 GHz LO chain. Similar trends can be observed for the ±1.33 GHz spurs which are generated in a manner similar to the spurs in the 20 GHz chain. However, additional spurs appear at ±666 MHz offset as well as at ±3×666 MHz offset. To explain the generation of the additional spurs, consider Fig. 5.32. The output of the BPF and ILFD is shown in terms of harmonics of the reference signal. When applied to the ILFD, the 11th harmonic is divided by two, generating a frequency corresponding to the 5.5th harmonic of the reference. The 10th and 12th harmonics are divided as well, generating very small components which fall onto the 5th and 6th harmonics of the reference. Due to the inevitable on-chip parasitic coupling (through electromagnetic radiation,

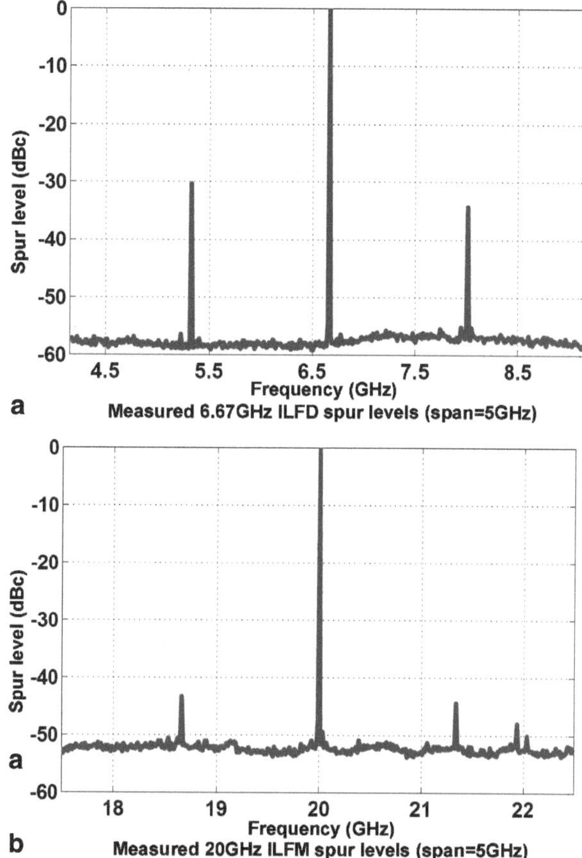

Fig. 5.30 Spurious performance of the 20 GHz LO chain. **a** Measured 6.67 GHz ILFD spur levels (sapn = 5 GHz), **b** Measured 20 GHz ILFM spur levels (sapn = 5 GHz)

substrate coupling, and capacitive coupling), the 5th and 6th harmonics at the output of the ILFD add to the similar components at the ILFD's input. A parasitic positive feedback loop is formed, enhancing these components which correspond to spurs at ±666 MHz from the desired 5.5th harmonic. The spurs at ±3×666 MHz are formed through third-order intermodulation of the ±1.33 GHz spurs and the ±666 MHz spurs. At the ILFM's output, the spur at +3×666 MHz is filtered out. The component at −3×666 MHz, however, remains and its amplitude increases due to parasitic reinforcement from the 20 GHz LO chain. One possible way to get rid of the ±666 MHz spur and the subsequent ±3×666 MHz spurs is to use a pulse slimmer that only generates odd harmonics in the 22 GHz LO chain. This further justifies the use of a separate pulse-slimmer for each chain as suggested in Sect. 5.3.5.

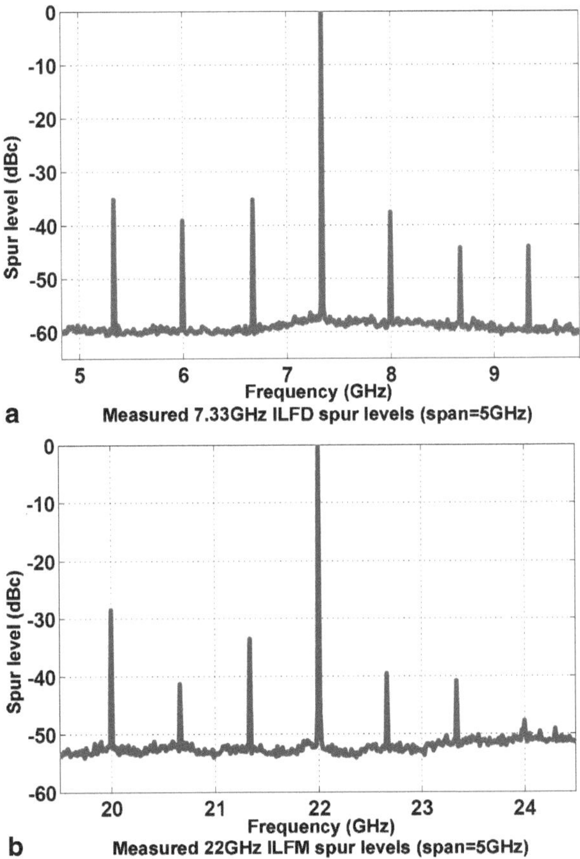

Fig. 5.31 Spurious performance of the 22 GHz LO chain. **a** Measured 7.33 GHz ILFD spur levels (sapn = 5 GHz), **b** Measured 22 GHz ILFM spur levels (sapn = 5 GHz)

The downconverted time domain I and Q outputs of the 20 and 22 GHz LOs are shown in Figs. 5.33 and 5.34 respectively. Due to a limitation of the off-chip balun, the output frequency cannot be below 100 MHz. This, in turn, means that the mismatches between I and Q paths on board are not negligible and add considerably to the measured phase difference. Nevertheless, the measured average phase error under these conditions is 10° and 2° for the 20 GHz and the 22 GHz LOs respectively.

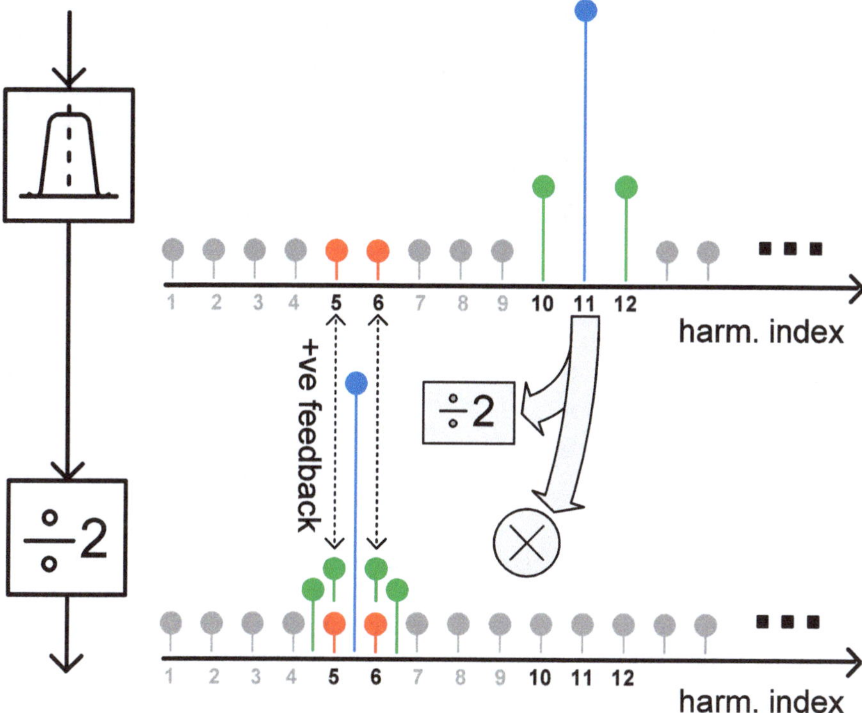

Fig. 5.32 Generation of ±666 MHz spurs in 22 GHz LO chain

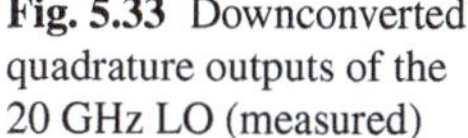

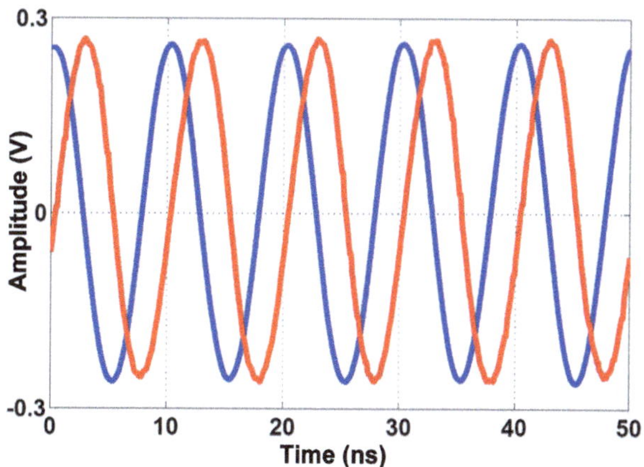

Fig. 5.33 Downconverted quadrature outputs of the 20 GHz LO (measured)

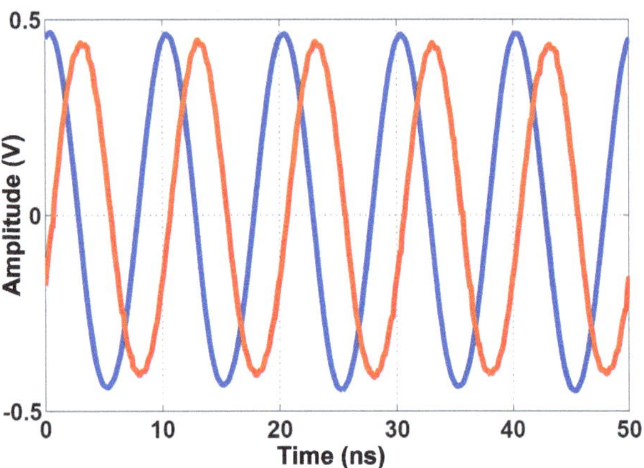

Fig. 5.34 Downconverted quadrature outputs of the 22 GHz LO (measured)

5.6 Conclusions

In this chapter an LO scheme for generating simultaneous phase synchronous quadrature LOs was presented for a 19–23 GHz channelized receiver. The architecture developed here can easily be extended to more than two channels, and can be implemented at different frequencies. This provides a viable mechanism to realize mm-wave channelized receivers. A number of novel techniques are used in the design, including a differentiating pulse-slimmer that exploits even and odd harmonics, an injection-locking based bandpass filter and an injection-locked quadrature frequency divider with direct injection.

Chapter 6
Conclusions

With the quest for software-defined-radio, frequency channelization holds the potential of realizing a system that is as close as possible to Mitola's architecture [5]. The major premise of Mitola's architecture is implementing an ADC with a bandwidth wide enough to accomodate the entire RF bands of interest. With current technology, direct implementation of such an ADC is an impossible feat. Nevertheless, proper extension of frequency channelization schemes can potentially achieve an affordable realization of the SDR. This venue of research includes finding the proper bandwidth to strike a balance between the overhead of the extra RF power needed for channelization and the total power consumption of the reduced sampling rate ADCs. Significant research is also required for multiple, simultaneous LO and clock generation schemes for channelization, as well as the potential issues of cross-talk and frequency pulling resulting from implementing multiple LOs on the same chip.

While frequency-channelization can bring the far-fetched Mitola's architecture closer to reality, the current approach to SDR involves relatively smaller instantaneous bandwidths transcievers, with tunable LOs. Pertinent to this SDR realization is the use of wide-tuning range QVCOs [6]. The ideas and schemes presented in this work can be built upon to enable robust quadrature LO generation, with the wider tuning ranges enabling a smaller number of QVCOs hence reducing the area requirements as well as the design and calibration complexity. Furthermore, mm-Wave SDR design remains an unexplored territory, and wide-tuning range mm-Wave QVCO's can present a starting point for such research, further extending the possible scope of applications of SDR.

M. Elbadry, R. Harjani, *Quadrature Frequency Generation for Wideband Wireless Applications,* Analog Circuits and Signal Processing, DOI 10.1007/978-3-319-13788-9_6

References

1. Evolution of cell phone to smartphone. https://blog.lookout.com/blog/2011/02/03/evolution-of-cell-phone-to-smartphone/. Accessed 20 Feb 2015.
2. Will LTE slap WiMAX and become the next-gen technology? http://www.broadbandwatcher.co.uk/will-lte-slap-wimax-and-become-the-next-gen-technology-566/. Accessed 15 Mar 2013.
3. Wirelesshd specifications. http://www.wirelesshd.org. Accessed 20 Feb 2015.
4. Namgoong W. A channleized digital ultrawideband receiver. IEEE Trans Wireless Commun. 2003;2(3):502–10.
5. Mitola J. The software radio architecture. IEEE Commun Mag. 1995; 33(5):26–38.
6. Bagheri R, Mirzaei A, Chehrazi S, Heidari ME, Lee M, Mikhemar M, Tang W, Abidi AA. A 800-MHz-6-GHz software-defined wireless receiver in 90-nm CMOS. IEEE J Solid-State Circuits. 2006;41(12):2860–76.
7. Razavi B. RF microelectronics. New Jersey: Prentice Hall; 2012.
8. Chen PW, Lin TY, Kei L-W, Yu R, Tsai M-D, Yeh C-W, Lee Y-B, Tzang B, Chen Y-H, Huang S-J, Lin Y-H, Dehng G-K. A 0.13 μm CMOS quad-band GSM/GPRS/EDGE RF transceiver using a low-noise fractional-N frequency synthesizer and direct-conversion architecture. IEEE J Solid-State Circuits. 2009;44(5):1454–63.
9. Siligaris A, Richard O, Martineau B, Mounet C, Chaix F, Ferragut R, Dehos C, Lanteri J, Dusspot L, Yamamoto SD, Pilard R, Busson P, Catheline A, Belot D, Vincent P. A 65-nm CMOS fully integrated transceiver module for 60-GHz wireless HD applications. IEEE J Solid-State Circuits. 2011;46(12):3005–17.
10. Murmann B. ADC performance survey 1997–2011. http://www.stanford.edu/ murmann/adcsurvey.html. Accessed June 2011.
11. Medi A, Namgoong W. A high data-rate energy-efficient interference-tolerant fully integrated CMOS frequency channelized UWB transceiver for impulse radio. IEEE J Solid-State Circuits. 2008;4(43):974–80.
12. Andreani P, Bonfanti A, Romano L, Samori C. Analysis and design of a 1.8-GHz CMOS LC quadrature VCO. IEEE J Solid-State Circuits. 2002;37(12):1737–47.
13. Behbahani F, Kishigami Y, Leete J, Abidi AA. CMOS mixers and polyphase filters for large image rejection. IEEE J Solid-State Circuits. 2001;36(6):873–87.
14. Rofougaran A, Rael J, Rofougaran M, Abidi A. A 900MHz CMOS LC-oscillator with quadrature outputs. In: International Solid-State Circuits Conference, Feb. 1996, pp. 392–3.
15. Rofougaran A, Change G, Rael JJ, Change JYC, Rofougaran M, Chang PJ, Djafari M, Ku MK, Roth EW, Abidi AA, Samueli H. A single chip 900-MHz spread-spectrum wireless transceivers in 1-μm CMOS —part I: architecture and transmitter design. IEEE J Solid-State Circuits. 1998;33(4):515–34.
16. Lanka NR, Patnaik SA, Harjani RA. Frequency-hopped quadrature frequency synthesizer in 0.13μm technology. IEEE J Solid-State Circuits. 2011;46(9):2021–32.
17. Razavi B. Design of millimter-wave CMOS radios: a tutorial. IEEE J Solid-State Circuits. 2009;56(1):4–16.

M. Elbadry, R. Harjani, *Quadrature Frequency Generation for Wideband Wireless Applications,* Analog Circuits and Signal Processing, DOI 10.1007/978-3-319-13788-9

18. Hajimiri A, Lee TH. A general theory of phase noise in electrical oscillators. IEEE J Solid-State Circuits. 1998;33(2):179–94.
19. Razavi B. A study of injection locking and pulling in oscillators. IEEE J Solid-State Circuits. 2004;39(9):1415–24.
20. Huang D, Li W, Zhou J, Li N, Chen J. A frequency synthesizer with optimally coupled QVCO and harmonic-rejection SSBmixer for multi-standard wireless receiver. IEEE J Solid-State Circuits. 2011;46(6):1307–20.
21. Elbadry M, Harjani R. Quadrature frequency synthesis for Wideband wireless transceivers. PhD thesis, University of Minnesota, May 2014.
22. Valla M, Montagna G, Castello R, Tonietto R, Bietti I. A 72-mW CMOS 802.11a direct conversion front-end with 3.5-db NF and 200-kHz $1/f$ noise corner. IEEE J Solid-State Circuits. 2005;40(4):970–7.
23. Vancorenland P, Steyaert MSJ. A 1.57-GHz fully integrated very low-phase-noise quadrature VCO. IEEE J Solid-State Circuits. 2002;37(5):653–6.
24. Yi X, Boon CC, Liu H, Lin JF, Lim WM. A 57.9-to-68.3 GHz 24.6 mW frequency synthesizer with in-phase injection-coupled QVCO in 65 nm CMOS technology. IEEE J Solid-State Circuits. 2014;49(2):347–59.
25. Gierkink SLJ, Laventino S, Frye RC, Samori C, Boccuzzi V. A low-pase-noise 5-GHz CMOS quadrature VCO using superharmonic coupling. IEEE J Solid-State Circuits. 2003;38(7):1148–54.
26. Chamas IR, Raman S. Analysis and design of a CMOS phase-tunable injection-coupled LC quadrature VCO. (PTIC-QVCO). IEEE J Solid-State Circuits. 2009;44(3):784–96.
27. Guermandi D, Tortori P, Franchi E, Gnudi A. A 0.83-2.5-GHz continuously tunable quadrature VCO. IEEE J Solid-State Circuits. 2005;40(12):2620–27.
28. Andreani P, Wang X. On the phase-noise and phase-error performances of multiphase LC CMOS VCOs. IEEE J Solid-State Circuits. 2004;39(11):1883–93.
29. Kim HR, C YC, S MOh, Yang MS, and SGLee. A very low-power quadrature VCO with back-gate coupling. IEEE J Solid-State Circuits. 2004;39(6):952–5.
30. Ng AWL, Luong HC. A 1-V 17-GHz 5-mW CMOS quadrature VCO based on transformer coupling. IEEE J Solid-State Circuits. 2007;42(9):1933–41.
31. Decanis U, Ghilioni A, Monaco E, Mazzanti A, Svelto F. A low-noise quadrature VCO based on magnetically coupled resonators and a wideband frequency divider at millimeter waves. IEEE J Solid-State Circuits. 2011;46(12):2943–55.
32. Hegazi E, Sjoland H, Abidi AA. A filtering technique to lower LC oscillator phase noise. IEEE J Solid-State Circuits. 2001;36(12):1921–30.
33. Ham D, Hajimiri A. Concepts and methods in optimization of integrated LC VCOs. IEEE J Solid-State Circuits. 2001;36(6):896–09.
34. Wu Q, Quach T, Mattamana A, Elabd S, Dooley SR, McCue JJ, Orlando PL, Creech GL, Khalil W. A 10 mW 37.8 GHz current-redistribution BiCMOS VCO with an average FOMT of -193.5 dBc/Hz. In IEEE International Solid State Circuits Conference, pp 150–51, Feb. 2013.
35. Cusmai G, Repossi M, Albasini G, Mazzanti A, Svelto F. A magnetically tuned quadrature oscillator. IEEE J Solid-State Circuits. 2007;42(12):2870–7.
36. Hong-Yeh C, Yuan-Ta C. K -band cmos differential and quadrature voltage-controlled oscillators for low phase-noise and low-power applications. IEEE Trans Microwave Theory Tech. 2012;60(1):46–59.
37. Wireless LAN at 60 GHz - IEEE 802.11ad Explained. Technical report, Agilent Technologies.
38. Wells J. Faster than fiber: the future of multi-G/s wireless. IEEE Microwave Mag. 2009;10(3):104–12.
39. Hasch J, Topak E, Schnabel R, Zwick T, Weigel R, Waldschmidt C. Millimeter-wave technology for automotive radar sensors in the 77 GHz frequency band. IEEE Trans Microwave Theory Tech. 2012;60(3):845–60.

40. Tabesh M, Chen J, Marcu C, Kong L, Kang S, Niknejad AM, Alon E. A 65 nm CMOS 4-element sub-34 mw/element 60 GHz phased-array transceiver. IEEE J of Solid State Circuits. 2011;46(12):3018–32.
41. Vidojkovic V, Mangraviti G, Khalaf K, Szortyka V, Vaesen K, Thillo W Van, Parvais B, Libois M, Thijs S, Long JR, Soens C, Wambacq P. A low-power 57-to-66 GHz transceiver in 40 nm LP CMOS with -17 dB EVM at 7 Gb/s. In IEEE International Solid State Circuits Conference, pp. 268–70, Feb. 2012.
42. Okada K, Matsushita K, Bunsen K, Murakami R, Musa A, Sato T, Asada H, Takayama N, Li N, Ito S, Chaivipas W, Minami R, Matsuzawa A. A 60 GHz 16QAM/8PSK/QPSK/BPSK direct-conversion transceiver for IEEE 802.15.3c. In IEEE International Solid-State Circuits Conference, pp. 160–2, Feb. 2011.
43. Kim C, Park P, Kim D-Y, Park K-H, Park M, Cho M-K, Lee SJ, Kim J-G, Eo YS, Park J, Baek D, Oh J-T, Hong S, Yu H-K. A CMOS centric 77 GHz automotive radar architecture. In IEEE Radio Frequency Integrated Circuits symposium, pp. 131–4, Jun. 2012.
44. Marcu C, Chowdhury D, Thakkar C, Park J-D, Kong L-K, Tabesh M, Wang Y, Afshar B, Gupta A, Arbabian A, Gambini S, Zamani R, Alon E, Niknejad AM. A 90 nm CMOS low-power 60 GHz transceiver with integrated baseband circuitry. IEEE J Solid State Circuits. 2009;44(12):3434–47.
45. Chan WL, Long JR. A 60-GHz band 2 × 2 phased-array transmitter in 65-nm CMOS. IEEE J Solid State Circuits. 2010;45(12):2682–95.
46. Scheir K, Bronckers S, Borremans J, Wambacq P, Rolain Y. A 52 GHz phased-array receiver front-end in 90 nm digital CMOS. IEEE J Solid State Circuits. 2008;43(12):2651–59.
47. Scheir K, Vandersteen G, Rolain Y, Wambacq P. A 57-to-66 GHz quadrature PLL in 45 nm digital CMOS. In IEEE International Solid State Circuits Conference, pp 494–5, 2009.
48. Wu L, Luong HC. A 49-to-62 GHz CMOS quadrature VCO with bimodal enhanced magnetic tuning. In IEEE European Solid State Circuits Conference, pp. 297–300, Sep. 2012.
49. Mazzanti A, Andreani P. Class-C harmonic CMOS VCOs, with a general result on phase noise. IEEE J Solid State Circuits. 2008;43(12):2716–29.
50. Integrand Software, Inc. EMX user's manual, 2010.
51. Sadhu B, Harjani R. Capacitor bank design for wide tuning range LC VCOs: 850 MHz-7.1 GHz (157 %). In Proceedings of 2010 IEEE International Symposium on Circuits and Systems (ISCAS), pp. 1975–78, 2010.
52. Shannon CE. Communication in the presence of noise. Proc IRE. 1949;37(1):10–21.
53. Goldsmith A. Wireless communications. Cambridge University Press, 2005.
54. Sai-Wang T, Socher E, Wong A, Yu W, Lan DV, Chang M-CF. Simultaneous sub-harmonic injection-locked mm-wave frequency generators for multi-band communications in CMOS. In IEEE Radio Frequency Integrated Circuits Symposium, pp. 131–4, Jun. 2008.
55. Chan WL, Long JR. A 60-GHz band 2 x 2 phased-array transmitter in 65-nm CMOS. IEEE J Solid-State Circuits. 2010;45(12):2682–95.
56. Abdul-Latif MM, Elsayed MM, Sanchez-Sinencio E. A wideband millimeter-wave frequency synthesis architecture using multi-order harmonic-synthesis and variable N-push frequency multiplication. IEEE J Solid-State Circuits. 2011;46(6):1265–83.
57. Elbadry M, Sadhu B, Qiu J, Harjani R. Dual channel injection-locked quadrature LO generation for a 4 GHz instantaneous bandwidth receiver at 21GHz center frequency. IEEE Trans Microwave Theory Tech. 2013;61(3):1186–99.
58. Dal Toso S, Bevilacqua A, Tiebout M, Marsili S, Sandner C, Gerosa A, Neviani A. UWB fast-hopping frequency generation based on sub-harmonic injection locking. IEEE J Solid-State Circuits. 2008;43(12):2844–52.
59. Izad MM, Heng CH. A pulse shapping technique for spur suppression in injection-locked synthesizers. IEEE J Solid-State Circuits. 2012;47(3):652–64.
60. Lathi BP, Ding Z. Modern digital and analog communication systems. Oxford University Press, 2009.

61. Kalia S, Elbadry M, Sadhu B, Patnaik S, Qiu J, Harjani R. A simple, unified phase noise model for injection-locked oscillators. In IEEE Radio Frequency Integrated Circuits Symposium, pp. 1–4, Jun. 2011.
62. Mazzanti A, Svelto F, Andreani P. On the amplitude and phase errors of quadrature LC-tank CMOS oscillators. IEEE J Solid-State Circuits. 2006;41(6):1305–13.
63. Kinget P, Melville R, Long D, Gopinathan V. An injection-locking scheme for precision quadrature generation. IEEE J Solid-State Circuits. 2002;37(7):845–51.
64. Chan WL, Long JR. A 56–65 GHz injection-locked frequency tripler with quadrature outputs in 90-nm CMOS. IEEE J Solid-State Circuits. 2008;43(12):2739–46.
65. Sadhu B, Ferriss MA, Natarajan AS, Yaldez S, Plouchart JO, Rylyakov AV, Garcia AV, Parker BD, Babakhani A, Reynolds S, Xin L, Pileggi L, Harjani R, Tierno J, Friedman D. A linearized, low phase noise VCO based 25 GHz PLL with automatic biasing. IEEE J Solid-State Circuits, 2012.
66. Mazzanti A, Uggetti P, Svelto F. Analysis and design of injection-locked LC dividers for quadrature generation. IEEE J Solid-State Circuits. 2004;39(9):1425–33.
67. Tiebout M. A CMOS direct injection-locked oscillator topology as high-frequency low-power frequency divider. IEEE J Solid-State Circuits. 2004;39(7):1170–74.
68. Andreani P. A 2 GHz, 17 % tuning range quadrature CMOS VCO with high figure-of-merit and 0.6° phase error. In IEEE European Solid-State Circuits Conference, pp. 815–8, 2002.
69. Craninckx J, Staeyart MSJ. A 1.8-GHz low-phase-noise CMOS VCO using optimized hollow spiral inductors. IEEE J Solid-State Circuits. 1997;32(5):736–44.
70. Kuhn WB, Ibrahim NM. Analysis of current crowding effects in multiturn spiral inductors. IEEE Trans Microwave Theory Tech. Jan. 2001.;49(1):31–38.
71. R&S FSP spectrum analyzer—data sheet.

Zeitfracht Medien GmbH
Ferdinand-Jühlke-Straße 7
99095 Erfurt, Deutschland
produktsicherheit@kolibri360.de